D-STAR FOR BEGINNERS 2ND EDITION

BRIAN SCHELL

D-STAR FOR BEGINNERS

Copyright 2015-2018 by Brian Schell. All rights reserved, including the right to reproduce this book or any portion of it in any form.

Written and designed by: Brian Schell
brian@brianschell.com

Version Date: July 18, 2018.
ISBN: 1511415096
ISBN-13: 978-1511415095
Printed in the USA of America

CONTENTS

Foreword to the Second Edition	1
What is D-Star?	3
Essentials of Using D-Star	9
D-Star with a Computer: DV-Dongle	21
Using D-Star with the Icom IC-92AD	41
Using D-Star with the Icom IC-ID5100A	59
Using D-Star with the DVAP	67
Raspberry Pi Interfacing	73
Going Forward with D-Star	75
Appendix A: D-Star Self-Registration	79
Appendix B: List of Reflectors	87
About the Author	95
Also by Brian Schell	97

FOREWORD TO THE SECOND EDITION

D-Star in itself hasn't changed since last year. It's still an immensely popular system for ham radio operators to talk to other hams worldwide without learning code or buying huge antennas. It's fun, it's social, and it's easy to use once it's set up. As the previous edition of the book demonstrated, it's not always easy to set a radio up for D-Star. Actually, it used to be a fairly complex procedure that made it difficult to get started. The latter half of 2014 saw the introduction of new Icom radios that vastly simplified programming and setup, making it much easier to get started.

Second Edition Notes:

The sections on the DV-Dongle and DVAP, other than minor editing, have not changed much. Other than the brief DVAP section on Raspberry Pi, there hasn't been much development with these two products.

The section on the Icom IC-92AD has been revised and

mostly rewritten for clarity and more examples. Sometime between the publication of the previous edition and this one, Icom has discontinued the 92AD. They are still widely available and very popular, so I have not removed the section; several of the older Icom radios use the same procedure for programming, so it still has value. Future editions will probably see at least the addition of one of the newer Icom handhelds.

I have added an entirely new section on the Icom 5100A radios. This far more modern radio is going to be the face of the future for D-Star. With a large touch screen, pre-programmed database, memories accessible by SD card, and included programming software, this new radio is quite a bit easier and more powerful than the previous generations.

And remember, the whole point of the book is to get you up and running quickly by eliminating the technical jargon and information you don't need to begin. There's a seemingly endless amount of technology and number of protocols and standards involved with the hardware and repeater side of D-Star, and we're going to ignore a lot of that here. The book isn't about setting up your own repeater; it's about setting up your personal radio or computer. Once you're connecting and talking on the repeaters, reflectors, and with other D-Star users, you'll find that there's still a lot more you can research and learn about D-Star that isn't included here.

Enjoy learning it all, and I'll talk to you on D-Star.

Brian Schell, KD8OTD
    KD8OTD@gmail.com

WHAT IS D-STAR?

D-STAR (Digital Smart Technologies for Amateur Radio) is an FDMA and GMSK digital voice and data protocol specification developed in the late 1990s as the result of research by the Japan Amateur Radio League to investigate digital technologies for amateur radio. - Wikipedia

That's a mouthful. Basically, it's a way of merging amateur radio and the Internet to allow a form of digital communications.

The beauty of D-Star is that it allows this communications in several ways:

1. **Radio Only**. You can use your handheld radio to connect with a local repeater that is D-Star enabled, and then use that link to talk to people elsewhere in the world.
2. **DVAP**. With a DVAP, you can plug a little dongle into your computer. Then you can use your own local handheld or base radio to transmit to the

computer, which takes your voice and transfers it over the Internet to other repeaters, reflectors, and users. This is essentially the same thing as the previous option, but this works if there is no local repeater in your area.
3. **DV-Dongle**. Another very popular use of D-Star is to just plug a dongle and a microphone into your PC and talk to hams around the world through the Internet. This is very similar to using Skype or another VOIP (Voice Over Internet Protocol) system, as no radio is required.
4. Depending on what hardware you have, you may not be able to use all the information in this book, but at least skim everything, even if you don't have the exact equipment the chapter discusses. We start out explaining the basic way D-Star and D-Star programming works in general. Then we move on to cover the DV-Dongle approach to D-Star, and then we talk about programming two specific example radios to use with D-Star. Lastly, we get into using the DVAP to combine the other approaches.

It's extremely flexible, and the sheer number of options is what tends to overwhelm people. Depending on the age of your D-Star radio, it can be either fairly complex or relatively easy to set up D-Star on your radio the first time. After it's all done in the beginning, then it's just a matter of how far you want to go with it.

The first major requirement is that in order to use the D-Star system, you *must* be a licensed amateur radio operator (ham). If you aren't a ham radio operator, then that's the necessary first step. In the USA, there are three main "levels" of license: Technician, General, and Amateur Extra. Technician is the beginning level, and it's not that hard to attain. The Technician license is all you need to do everything D-Star offers. The higher-level licenses will allow you to do other fun things with your radio, but they aren't necessary at all to get all the benefits of D-Star.

But I'm Not a Ham!

If you aren't already a licensed amateur radio operator, you will need to become one to proceed. A great place to start if you aren't already licensed is the **ARRL Ham Radio License Manual**. It's the book that gets most hams started. With that out of the way, everything else in this book will assume you *are* a licensed amateur. Again, it doesn't really matter which level of license you have for D-Star, just get licensed.

Is D-Star Really "Radio?"

There has always been controversy over D-Star by purists who insist that it "isn't real radio." There's really no reason to debate this, it's really just a matter of how you use it. If you are holding a handheld radio and using it to connect to local repeater that connects through the Internet to a repeater in Sweden that is being used by a Swedish ham who is also holding a radio in his hand, it's hard to argue that this isn't "real radio." It's "Internet-assisted" radio, but there are still actual radios involved on both ends.

On the other hand, if you are sitting at your desk, talking into a microphone connected to your computer and D-Star Dongle, with no radio in sight, and you are connected through D-Star to a user in Toronto who is running the same setup on his computer, is that "real radio?" Probably not, since there's no over-the-air radio transmission involved anywhere in the process. It's really no different from talking on Skype.

The reason D-Star users need to be licensed hams is that *you don't know* what the person on the other end of the connection is using. If you are on the computer and connecting to a ham in Sydney Australia, you don't know if he or she is on a cheap old PC or sitting in front of a radio system that cost more than your car. You just can't tell, so the licensing requirement is there to cover all the possibilities.

What Can I do with D-Star?

There are numerous benefits to D-Star. First and foremost, it's all-digital. With a good connection, this means you have crystal-clear audio and sharp voices all around. Or on a less-than-optimal connection, you get nothing but noise, often called "R2-D2" after the incomprehensible sounds of the famous droid from Star Wars. There are no half-heard or weak signals with D-Star; it's all or nothing.

The next benefit is the flexibility. You can connect all sorts of different ways. You can link to repeaters thousands of miles away and call CQ on a local repeater in Germany, or Australia, or New Jersey. You can link to a specific user's radio without knowing where in the world he's located. You can connect to large "chat room" areas called reflectors and find someone to talk to almost any time of the day. There is even software called D-RATS, that allows you to send data files and text messages through D-Star.

It's not a perfect system, and it has its detractors, but then there are still hams out there who claim anything other than Morse code isn't "real radio." Ham radio is a big hobby, with lots of avenues to explore, and D-Star is only one of them.

First Step: Registration

The absolute first step, regardless of how you plan to connect to D-Star, is to get yourself registered on the D-Star network. See **Appendix A** on how to do that. It may take a few days to get approved, so it's probably a good idea to do this right after you've ordered your radio or Dongle, so your registration will be ready to go when you get it. It doesn't matter which method of connecting to D-Star you plan to use, you need to be registered in advance for any of them, so do not skip this step.

Terminology

Throughout the book, I will talk a lot about repeaters and reflectors, and it's easy to get the two mixed up. They are not the same thing:

Repeater: A repeater is a device that sits high up, usually on the top of a tall building or tower somewhere. It accepts low-powered transmissions from nearby radios and re-transmits that signal through a more powerful radio signal and/or to somewhere else through the Internet.

Hams use repeaters all the time on the VHF and UHF bands.

Reflector: Reflectors are unique to D-Star. They are a lot like a party line or "chat room" where you can just talk to other hams without any concern for where they are. If you are familiar with Echolink, another system for connecting radios and the Internet, a reflector is similar to an Echolink Conference. Reflectors are named with a combination of a number and a letter (for modules, which we'll get into later). For example, REF001C is "Reflector one Charlie." At the time of this writing, there are 71 reflectors, each with several modules (A, B, C, and sometimes D).

ESSENTIALS OF USING D-STAR

Getting Started

Before we get started actually programming your radio, we need to see how D-Star handles commands and does things. This chapter explains the theory and process as generically as possible so you know what needs to be done and what information you will need to research on your own. The following several chapters are more hands-on in nature.

As previously stated, the absolute first step is to get yourself registered on the D-Star network, and Appendix A tells you how to do that.

The second step is to make sure you actually have a local repeater that supports D-Star. If you're in a major metropolitan area, you probably do, but there are still some surprising gaps in D-Star coverage, so it wouldn't hurt to check before you waste big money for a D-Star radio that you can't use. To find out which D-Star repeaters are in range and active in your local area. There's a really handy tool at:

http://www.dstarinfo.com/dstar-web-calculator.aspx

You choose your country, state, city, and possibly more than one repeater. If your city only has one repeater, then write down the information it gives you and move on. If there aren't any repeaters nearby, then that's information you need to take into account as well. If you're really lucky, you live in a city with multiple D-Star repeaters.

In my case, I can see from that site that both

"USA, Dayton, Ohio, **W8RTL** Gateway" and "USA, Dayton, Ohio, **W8HEQ** Gateway"

are in my town. I also drive through Cincinnati sometimes, and that will sometimes put me in range for

"USA, Cincinnati, Ohio, **K8BIG**"

I will want to eventually set up all three of them in my radio, so they are available when I want to use them. I have no way of knowing which of these three is going to work best for me at home, or which I will use the most, but it's simpler for now just to set up all three of them and figure out which one works best for me later. Go through the website list and make a note of every possible repeater that you might be able to connect to. Maybe your town only has one, or maybe you travel a lot. In my case, I only need to deal with those three.

Connection Information

There are four bits of information you will need to have in

order to connect to anything on D-Star. There are thousands of possible combinations of callsigns, modules, and other stuff, so the D-Star calculator I linked to above makes sorting all of it out vastly easier. Each one of these four fields is entered into the radio in **8-character** strings in order to do anything with D-Star.

- Originator (**MYCALL**) This is your callsign. Mine is KD8OTD, and that's what I'll be using in the following examples throughout the book. If you will be the only user of this radio, then you can "set it and forget it," as it won't change.
- Repeater1 (**RPT1**) This is the callsign of the repeater port you use to access the local repeater. I use W8RTL Module C, so that's what I'll be using below. If you're running a base radio, or otherwise don't travel often, then this is another one that won't change much.
- Repeater2 (**RPT2**) This is the callsign of the repeater port you use to connect the local repeater to somewhere else. Mine is W8RTL G ("G" is for Gateway).
- Destination (**URCALL** or **YOUR**) This is the callsign of the person, group, reflector, or repeater you want to talk to. Sometimes this field also has some kind of command string that causes the D-Star repeater to do something. We'll talk about this a lot more later on.

You can use the D-Star calculator website to "describe" what you want to do, and the site will give you all that you need to fill in these four fields. In the next section, I am going to walk through the D-Star calculator to set up W8RTL as my local repeater. Of course, you'll set up your own local repeater on your real radio, but for now I would recommend following along with me and looking up my repeater on the calculator site, just to see if you can get the same results I do. Here is that link again:

http://www.dstarinfo.com/dstar-web-calculator.aspx

The first step is to connect to my local repeater. The first one I'm going to look up is W8RTL. If you know which local repeater you want to set up first, then select it as your source repeater in the first line in the green box, as below:

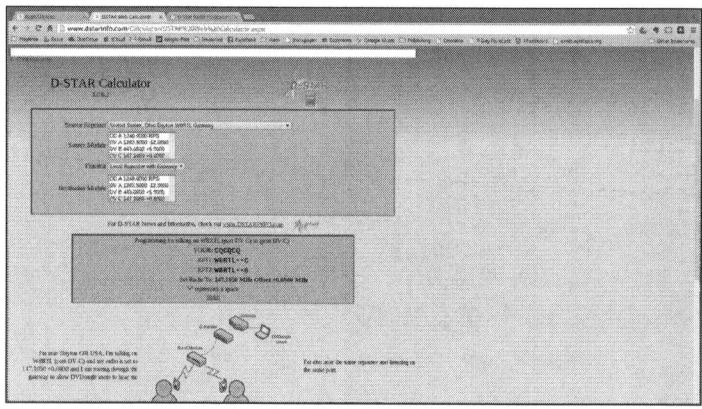

D-Star Calculator Website

Next you'll need to select a source and destination module. The screen above tells me that for this particular repeater, Module A is at 1283.5 mHz, Module B is at 443.05, and Module C is at 147.105. The exact frequencies will be different for each specific repeater, but notice that the A module is on the 23cm band (1.2 GHz), the B module is on the 70cm band (440-450 MHz), and the C module is on the

2m band (144-148 MHz). In the box for the "Source Module," you will need to pick whichever module your radio can connect to. For example, if your radio only supports 2m and 70cm, then you can't connect through the 23cm A module, but you could pick either the 2m C module or the 70cm B module. Which you choose is completely up to you, as long as your radio can transmit in the appropriate band.

Since my example radio is a 2-meter radio, I will choose the C module. Just to keep things simple, I will choose C for both Source and Destination.

You can choose whichever Destination module you want; you are not limited by your radio.

Look at the various options under the "Function" drop-down list. This is where most of the functionality of D-Star comes from. For now, select "Local Repeater with Gateway." Once you have selected a Source Repeater, Source Module, Function, and Destination Module, the screen will refresh, and the bottom portion of the screen will now show a blue box with some instructions:

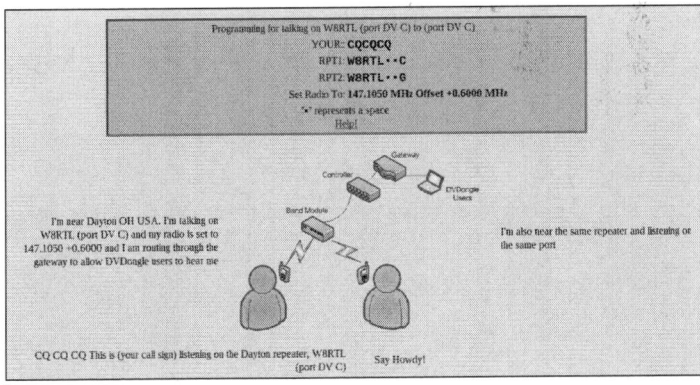

Local Repeater with Gateway Screen

Inside the blue box, it says:

```
Programming for talking on W8RTL (port DV
C) to (port DV C)
YOUR: CQCQCQ
RPT1: W8RTL■■C
RPT2: W8RTL■■G
Set Radio To:
147.1050 MHz Offset +0.6000 MHz
"■" represents a space
```

And below that, the little cartoon-men are having a conversation. If what the cartoon-men are saying sounds like what you intended, then you are doing great. If not, go back and change your settings in the green box as many times as you need to.

Notice that inside the RPT1 and RPT2 fields in the blue box there are placeholders that represent spaces. Each string that you enter into the radio to do anything with D-Star **must have eight characters** and be in a particular format. If the calculator site tells you there are spaces in a command string, then be sure to take note of that; it'll be important when actually programming your radio.

By this point, you should know the information for:

Repeater Name
Frequency
Offset
YOUR
RPT1
RPT2

Repeat this step as necessary to get connection information for **EACH repeater** you plan to program for the next step. If you have more than one local D-Star repeater that you want to use, you'll need to look up the information for each of them. If you have been following along with my W8RTL example, then clear that off and do it for your own repeaters. I'll wait!

Destination Programming

What you have right now is theoretically enough to connect to your local digital D-Star repeater and talk to people locally or to those who are using D-Star Dongles. That's nice, because it's digital, but it's not really going "outside" your local area. In order to move beyond your local region, you'll need to connect that D-Star repeater to some other destination.

Let's return to the D-Star calculator website. Look more closely at the "Function" drop-down. You'll see at least the following:

- Local Repeater
- Local Repeater with Gateway
- Echo Test
- Repeater Status
- Source Route
- Specific User
- Link to Reflector
- Link to Repeater
- Unlink Repeater
- High-Speed Data

In normal usage, you'll use the "Local Repeater with Gateway" function in the way I already demonstrated to learn how to access your local repeater. If you don't travel much, you can set this up and forget it; you'll always access the same the repeater the same way. The functions, on the other hand will change with every session. Here's a typical example flow of a D-Star QSO:

1. You get **INFORMATION** about the repeater to see if it's linked to something already. You may or may not need to **UNLINK** it from there if it's connected.
2. You then **LINK** to something new: A reflector, repeater, or another user.
3. You talk on that link.
4. You **UNLINK** the repeater when you're done.

Connect to a Reflector

Now it's time for another example. Let's say you want to listen in or participate in a fairly busy discussion. That's what Reflector 1C is like, so we'll do that (Appendix B has a list of reflectors and what they're most often used for). Going back to the D-Star calculator page, I plug in my local repeater (W8RTL), the source module that I want to use on my radio (C module for 2-meters), the function (LINK TO REFLECTOR), and the destination reflector (REF001C). If I plug in the information in the four source fields, it returns:

> Programming for linking from W8RTL (port DV C) to REF001 (port DV C)
> YOUR: **REF001CL**
> RPT1:   **W8RTL■■C**
> RPT2:   **W8RTL■■G**
> Set Radio To: **147.1050 MHz Offset +0.6000 MHz**
> "■" represents a space

Web calculator for Reflector 001C

And you'd need to program all that into your radio, which we'll cover in a later chapter. Before we get to that, let's look at these fields that the calculator gave us. With a little practice and experience, you won't need the calculator any more.

Programming for linking from W8RTL (port DV C) to REF001 (port DV C) This line explains what the calculator thinks we are trying to accomplish. Since we want to link our local repeater to Reflector 1C through the C module, everything looks correct with this line. If you don't get the results from your radio that you expect, go back and look at this line; it could be that the page misunderstood what you wanted to do.

YOUR: REF001CL The YOUR field is the function or command we are sending to the repeater. In this case, we are connecting to REF001, Module "C", and using the link command, which is a capital letter L. Notice that all of this combined is exactly 8 characters long.

RPT1: W8RTL ■ ■ C This is the local repeater and module C that we want our radio to connect to. Notice that there are two spaces between the C and the reflector's callsign. Again, this **must** add up to eight characters. And again, this module depends on what band you will use to connect to the repeater.

RPT2: W8RTL■■G Similar to the above, but this is the outgoing gateway of the repeater. The G is for Gateway, and it's used to connect, through the Internet, so some distant destination.

Set Radio To: 147.1050 MHz Offset +0.6000 MHz
These are the frequencies and other radio settings you will need to actually get your radio to communicate with the repeater.

Depending on what you want to do, here are other command strings in the YOUR field,

- **L = Link**
- **CL = Link to Module C (BL and AL are similar)**
- **I = Information about the repeater**
- **U = Unlink.**
- **E = Echo Test**

There are others, but they are rarely used. Experiment with the Function drop-down on the calculator site to see how the various functions translate into command strings.

Repeater Linking

Reflectors are a lot of fun, but with D-Star, you also have the option to connect to other D-Star compatible repeaters. If you know you will be connecting to a repeater in another city, you can program in the callsign of the repeater here. For example, if I wanted to talk through the K8BIG repeater in Cincinnati, I would add a line (under "YOUR") like this:

K8BIG*CL

Again the "*" is where you enter a space. This will (L)ink me to Module (C) of K8BIG.

You can once again use the D-Star calculator website to help you set this up. It works basically like the previous example, but instead of setting the Function to "Local Repeater with Gateway," you instead choose "Link to Repeater." That calculator site is your best friend; here it is again: **http://www.dstarinfo.com/dstar-web-calculator.aspx**

Now when you connect to that distant repeater, you can CQ or call anyone in that distant city through their local repeater. From here in Dayton, I can talk to hams all over the world this way.

Link to a Specific User

The above information explains the general procedure for linking to reflectors and/or distant repeaters. For many users, this is all that will be needed.

You can also use D-Star to connect to a specific, individual D-Star user in a long-distance conversation. Most of the setup is the same, you just set the "YOUR" field to the other person's callsign. If I wanted to call K8ABC and use my local W8RTL repeater, I'd set it up like this:

YOUR: K8ABC* (3 spaces after callsign)
RPT1: W8RTLC** (C=Module C again)
RPT2: W8RTLG** (G= Gateway again)

If K8ABC is listening, it will make a link between the two radios, wherever they happen to be. Please note that like all amateur radio, this is a public connection, and anyone listening to the repeaters on either end of the conversation will hear it all. For private conversations, use the telephone!

Play around with the D-Star calculator tool. Once you see how it works out everything you probably won't need it any more, but it makes learning how to setup all the many options for command strings a lot more foolproof.

OK, now we have researched all the various bits of information you will need in order to use your Dongle or radio, so let's actually **do** something now!

D-STAR WITH A COMPUTER: DV-DONGLE

The DV-Dongle is a small blue "dongle," or little plug-in device that looks something like a big blue flash drive with a cable. You can find out more information about the device at **http://www.dvdongle.com**. When a DV-Dongle is connected to a PC or Mac and used with the DVTool software, an amateur radio operator can connect to the international D-Star gateway network and receive and transmit just like a D-Star radio user.

If you don't have a D-Star repeater nearby, or you don't have a D-Star compatible radio, the DV-Dongle may be what you're looking for. These little devices cost around $200 new from most ham radio equipment retailers. Sometimes you can pick up used units very inexpensively from second hand stores or online sites like eBay. Less expensive devices that operate similarly are coming on the market now; more on that at the end of the chapter.

In the box, you get the blue DV-Dongle, a cable, and a "getting started" card. You'll need to supply your own computer and microphone.

The "Blue Dongle" - The DV-Dongle

1. Make sure your computer has the necessary hardware. You need an Internet connection. Your Internet connection doesn't have to be anything fast or outstanding, even old-fashioned dial-up is often fine for D-Star. That being said, coding and decoding the audio signal takes a good deal of processing power, so a fast computer is definitely helpful. A 2 GHz or higher processor with 512MB RAM is recommended. You'll also need either Windows XP or newer, Mac OS X Leopard or newer, or some modern flavor of Linux. **There is no way of which I am aware to use D-Star with an Apple or Android smartphone or tablet as of this writing.**
2. You'll need a microphone. There are a thousand varieties of microphone, but for D-Star you don't need anything especially expensive or super-high quality. Most laptop computers have a microphone

built-in, and these are *often* fine for D-Star. If you have a microphone of some kind already, then proceed with the rest of the installation. When we get to the "Testing" portion of the chapter, you'll find out whether you are happy with the quality or not. If you don't have a microphone, I can recommend the **Logitech ClearChat Comfort / USB Headset H390**, which includes a USB headset with the microphone. It's the one I use, and it works great for me.

That's all you need equipment-wise for a very basic setup. The very first step is to plug in the dongle into any available USB port. The little green light inside should glow if it's getting power. Also make sure your microphone and/or headset is plugged in.

The next step is to download the software. Go to **http://www.opendstar.org/tools/** and click the link to download the appropriate "DVTool" file for your system:

DVToolInstaller-2.0beta5.exe (Windows)
DVTool-2.0beta5-linux (Linux)
DVTool-2.0beta5-mac.dmg (Mac OSX)

I recommend trying the beta versions of the software first. The programmers have greatly improved the 2.0 version of the software over the original software offered at **http://dvdongle.com**, so use the newer version unless you have trouble. If something goes wrong, and it doesn't work, you can always go back and download an older version, but the beta 2.0 versions should work for most users.

The following walkthrough of the installation shows screen captures from the Windows version, but the Mac version is very similar.

This is the initial screen after clicking on the file you downloaded. Click on "Next."

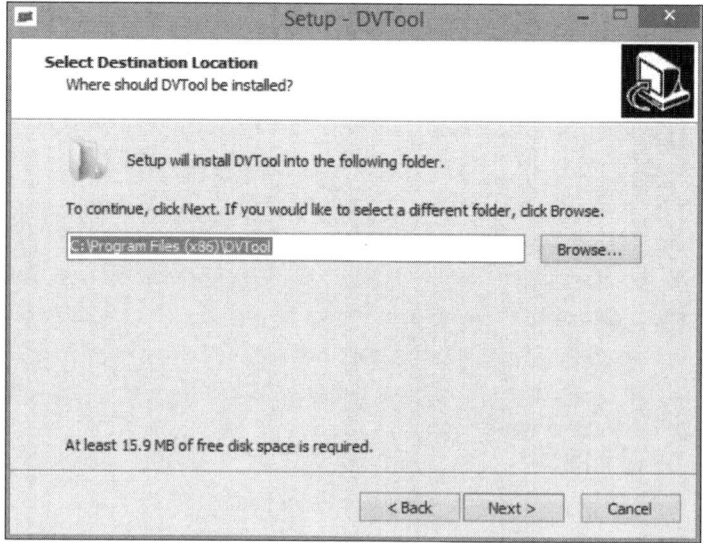

Here is the screen where you choose the destination of where to install the software. Unless you have a special reason to change it, just click on "Next."

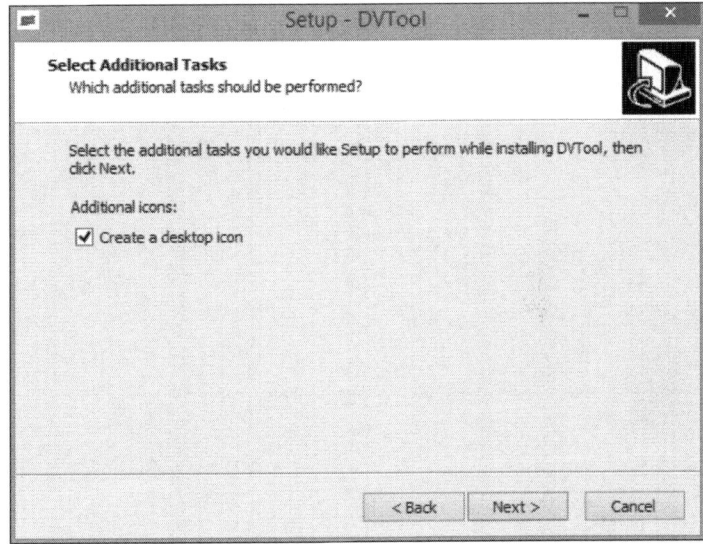

You can create a desktop icon or not, depending on your preference. I like having the shortcut on my desktop, so I chose this option, but you may prefer it otherwise.

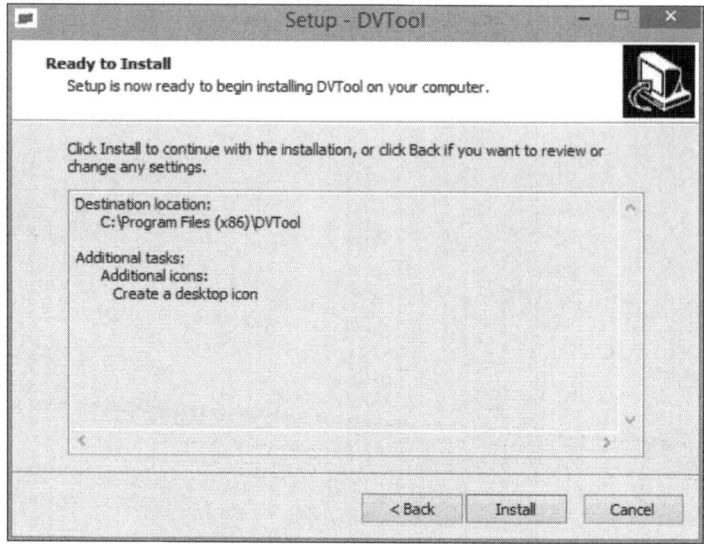

Last chance to change any of the install options. If this all looks good, then click "Install."

Click "Finish" to get it all installed. The DV Tool program should start automatically when the installation is complete.

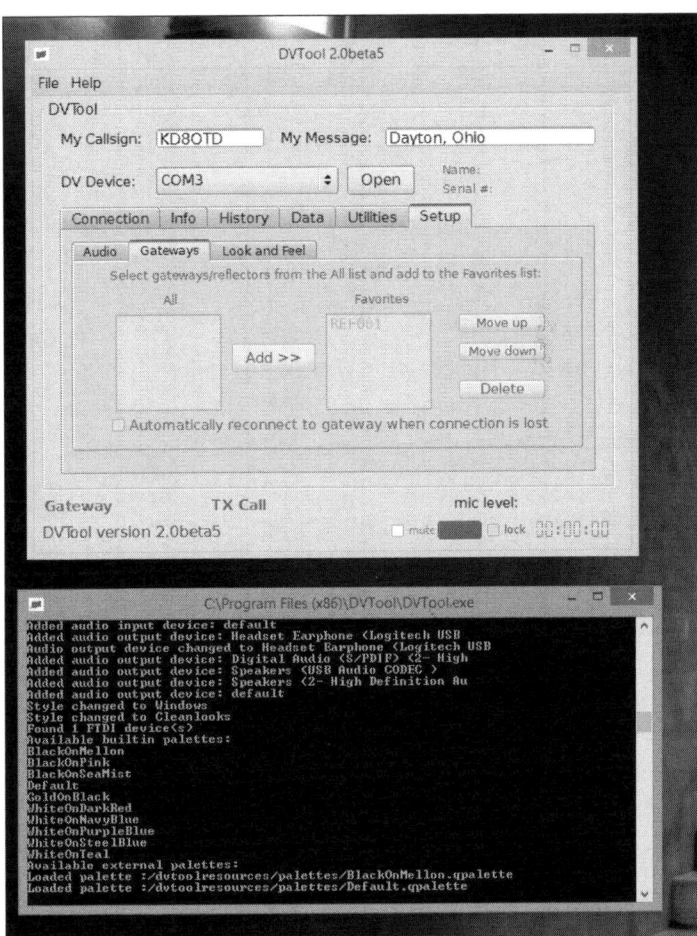

You'll notice that it opens two windows. One looks like a regular Windows program while the other is an all-text terminal view.

Most of the time, you don't need to see the terminal window, so you can minimize it to get it off your screen, but it always needs to be running. Do *not* click the red "X" to close it. If you start getting error messages or there is a problem with the network or audio stream somewhere, you can often track down the issue by looking at this terminal window. Generally though, if everything is working properly, you'll spend most of your time in the regular DVTool window. For now though, leave both on the screen so we can see what it does.

Let's take a look at what you need to set up on the main screen. First, you need to enter your amateur radio callsign. You'll need to enter a message. For a message, I usually put my location "Dayton, Ohio," but you can put any short message you like in there. It doesn't really matter.

Next, take a look at the "DV Device" dropdown. Depending on what other devices you have plugged into your computer, your software may not say COM3 like the screenshot. Pull down the dropdown box and choose the port that you used.

If you have more than one option and you aren't sure which to use, you have two choices.

1. Guess. If the rest of the steps don't work, change this setting and try again. Trial and error is the favorite solution for most hams!
2. Open up the Windows control panel (or preferences in OSX) and choose "Device Manager." Look at the options under "Ports (COM & LPT)" and see what's there:

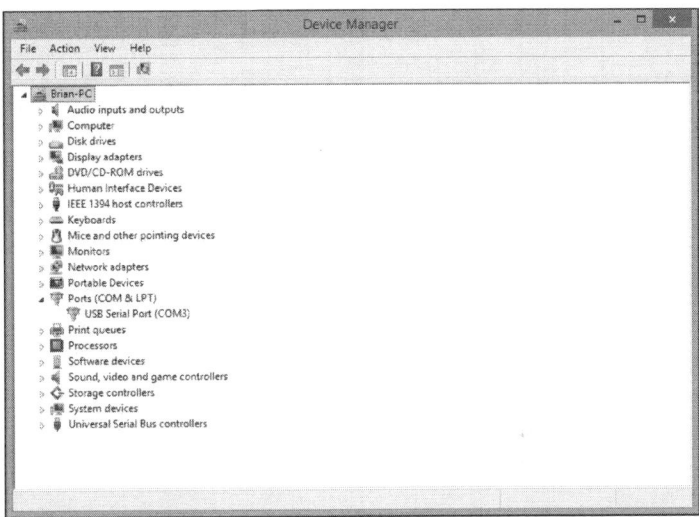

Windows Device Manager

In my case above, COM3 is the only option. Your system may show something else.

Once you've picked out your port, do not click on "Open" yet. We still have to set up the microphone and speakers. Click on the "Setup" tab and then choose the "Audio" sub-tab.

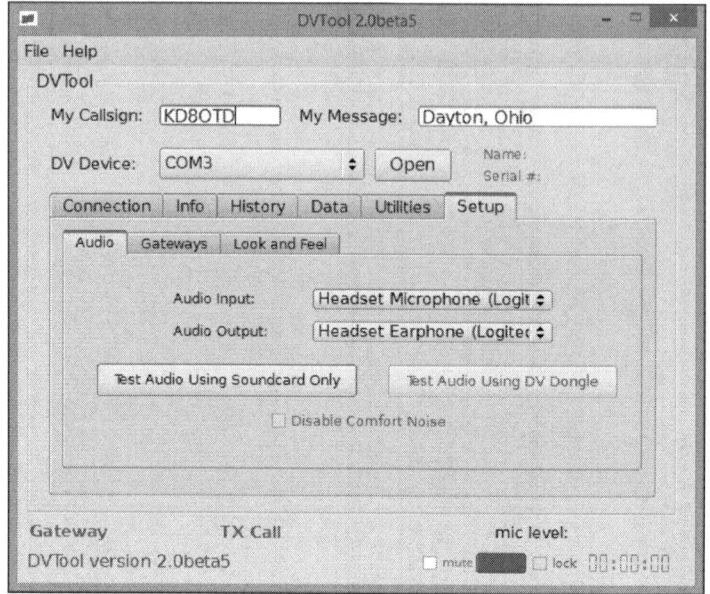

Choosing Audio Input and Output Devices

Use the "Audio Input" and "Audio Output" pull-down menus to choose which "Audio Input and Audio Output" you want to use. I use the Logitech headset that I mentioned earlier, and you can see those in the screenshot above. Once you have what you think are the correct options, click on "Test Audio Using Soundcard Only" and speak into your microphone. If you can hear yourself in the output device, then you're good to go. If you cannot hear your voice being repeated, then play with the input and output settings, and make sure your headset is plugged in and working properly. Be sure to check your sound volume.

If your test succeeded and your D-Star registration has gone through (See Appendix A), you are ready to try to connect to the D-Star Network. Click the "Open" button next to the DV Device you selected earlier. You should see some stuff scroll through the terminal window. That's a good thing. Your device is now live.

Next, click on the "Connection" tab. This is where you decide whom or where you are calling. Probably the busiest place on the D-Star network is Reflector 1C. Reflectors are something similar to ham radio repeaters, but are a D-Star only phenomenon. For our first connection, let's try connecting to Reflector 1, Module C.

Use the dropdown next to "Connect to Gateway" to select REF001. Notice that there are many reflectors on the list, followed by many amateur radio callsigns. These callsigns are repeaters scattered around the world; you can connect to any of them at this point. For now though, Choose REF001 and just to the right of that, choose "Module C."

Before we continue, it might be a good time to cover what those "Modules" are. We just connected to Module C on REF001.

Most D-Star Repeater installations have more than one actual repeater. Each of these repeaters is set up on a different amateur radio band. Typically they are set up like this:

The C module is on the 2 m band (144-148 MHz).
The B module is on the 70 cm band (440-450 MHz).
The A module is on the 23 cm band (1.2 GHz).

When using the Dongle, you don't need to worry about local frequencies, since you aren't connecting to a local repeater. In other cases, for example when connecting to a distant repeater, they may be very important. For example, if you are connecting to a repeater in Sweden, and you choose the A module, then only hams near that repeater who have radios capable of connecting on the 23 cm band will be able to respond. Maybe that's what you intended, maybe it's not, but it's something to consider.

Finally, click on "Connect to Gateway." If all goes well, you will see "Gateway REF001" appear at the bottom of the window, with REF001 in red.

If you're lucky, someone may be holding a conversation right now. Unless you see an error message in one of the two windows, wait and listen for a few minutes and see what happens.

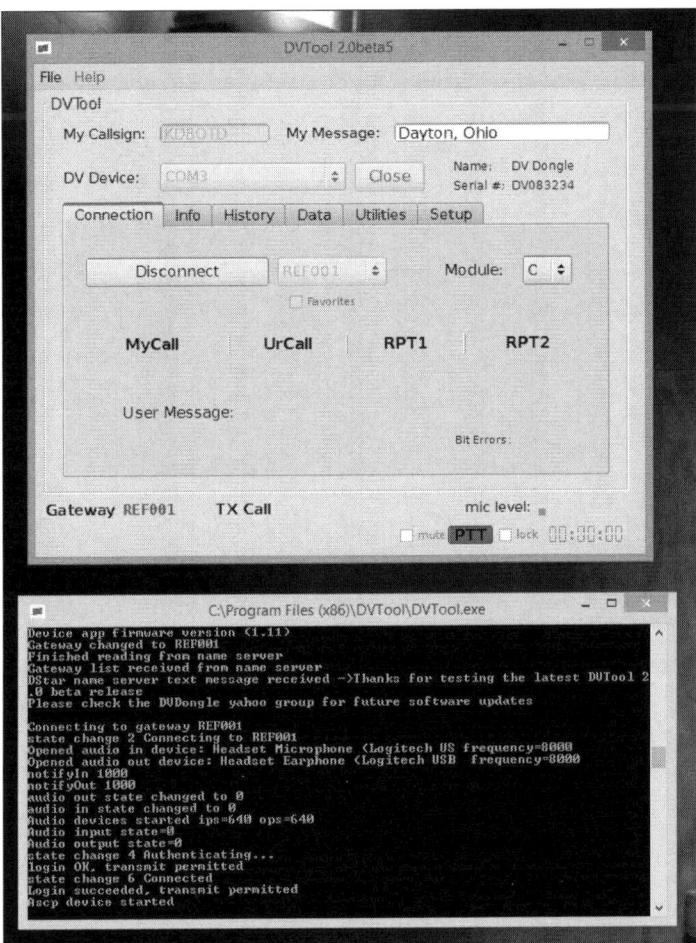

Notice the text towards the bottom of the screen in the terminal window. "Login succeeded, transmit permitted, etc." If you don't see this, then something, somewhere may be wrong.

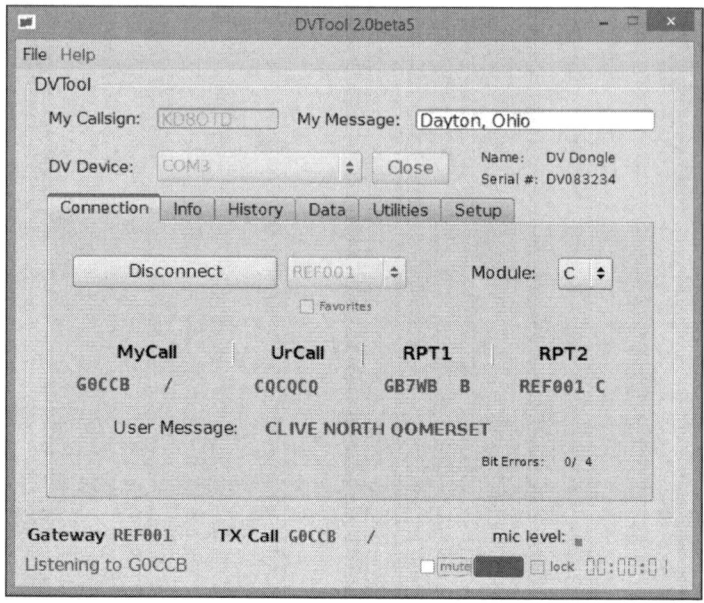

When someone else is talking, you'll see their callsign show up under "MyCall," usually "CQCQCQ" shows up under "UrCall" and the repeaters they are using show up under RPT1 and RPT2. Most of the time, their location or name or other message shows up under "User Message." In the screenshot above, G0CCB is talking through the B module on his local repeater (GB7WB), and I hear him on REF001 module C. The message he's posted "CLIVE NORTH QOMERSET" indicates his name is Clive and he's from North Qomerset.

If the conversation you are listening to is over, or if there is no one on right now, try calling CQ and see if someone responds. At the bottom of the screen is a green button labeled "PTT" (Push To Talk). Click it with your mouse and call CQ just like you would with a radio. "CQCQ, this is KD8OTD, Brian from Dayton, Ohio," and see what develops.

You can either click the PTT with your mouse or just hold down the space bar.

Notice the color of the PTT button. If no one else is speaking, it will be dark green. If someone else is speaking, it will be red or gray and you can't press it. This is a convenient way to keep people from trying to talk over each other.

After you've listened and talked for a bit, let's look at some other parts of the software. Click on the "Info" tab.

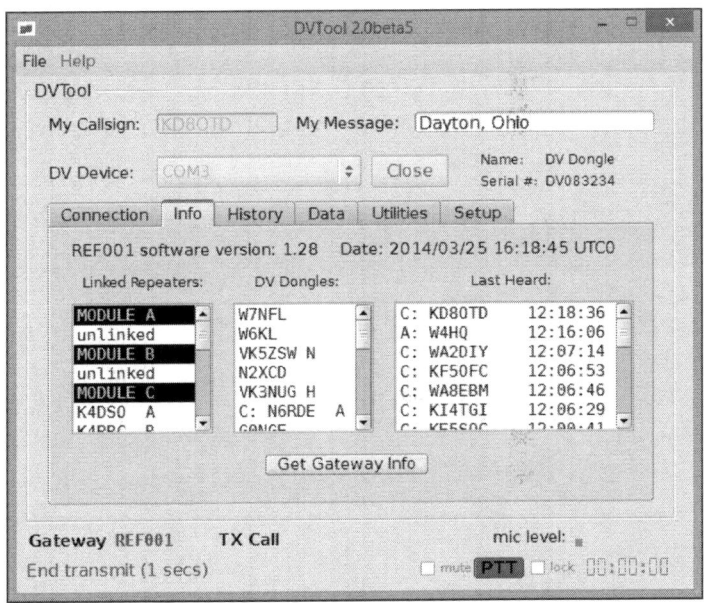

Click on "Get Gateway Info" and the three main boxes will populate with information. You can see all the repeaters that are connected to this reflector as well as a list of all the other DV-Dongle users that are listening. You can also see who the last few people who talked on that reflector were. In the screen capture above, I was the last person to talk on Module C at 12:18 p.m.

Now click on the "History" tab.

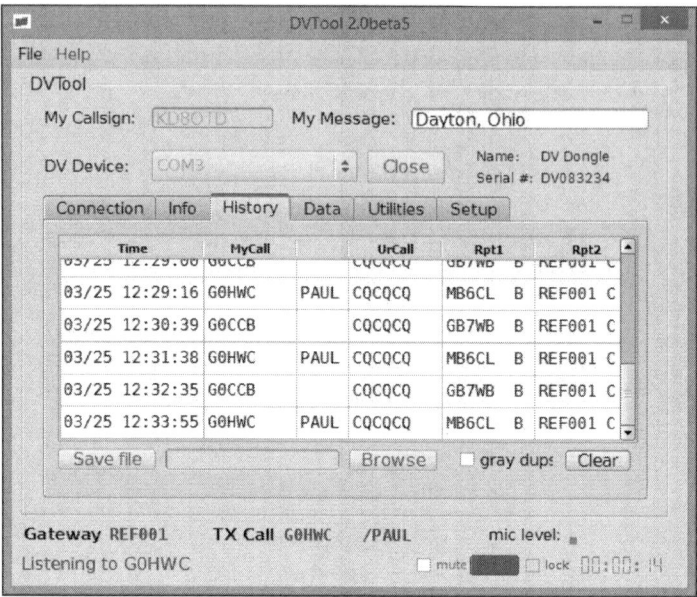

This screen simply lists a new line every time someone hits the PTT button. What we're looking at above is a conversation between G0HWC and G0CCB. You can see they both go back and forth a couple of times, which is normal in a conversation. You can scroll back through the list as far back as you want to see if someone has been on recently.

Let's take a look at the "Data" tab next.

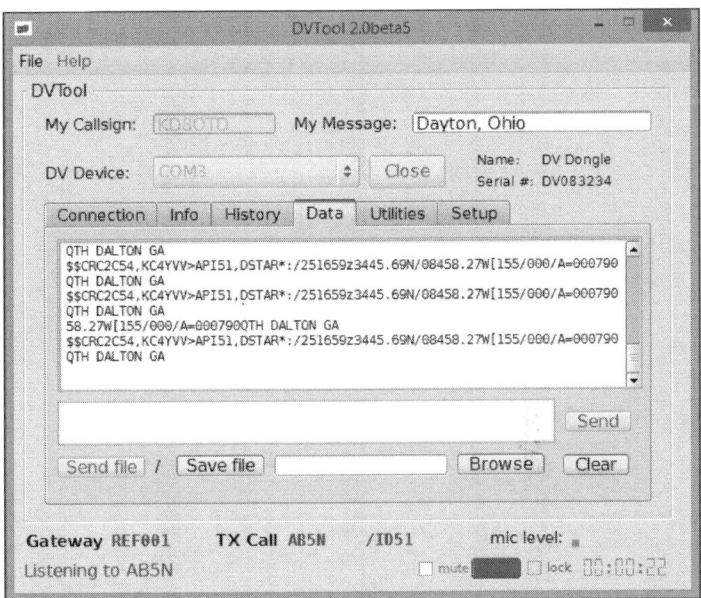

These are the raw data packets that accompany each audio transmission.

NW Digital Radio's DV-Thumb

The blue DV-Dongle was the first dongle of its type to be made available, and for years was the only non-radio option. Not any more, though. A new dongle similar to the DV-Dongle is the DV-Thumb. Created and sold by NW Digital Radio, the DV-Thumb is another dongle solution that lets you connect to the D-Star network through your computer. It is essentially a less-expensive version of the DV-Dongle. It's not as mature of a product, and the software is not as stable, but at $119.95, it's far less expensive than the original Dongle.

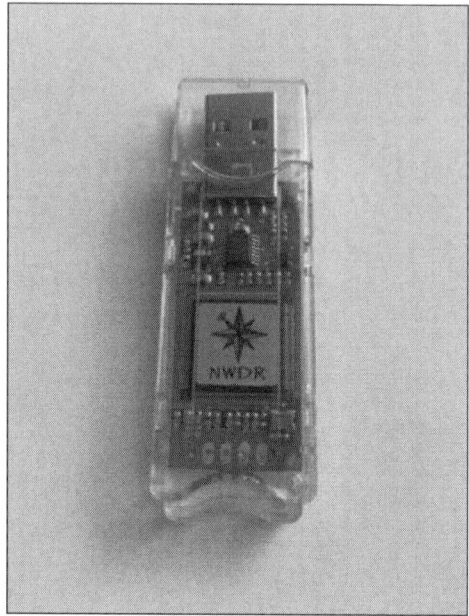

DV-Thumb

If you are adventurous, give it a try, and take part in the open-source software development for the product. If you're really into the computer-side of things, they also sell the DV3000, a small circuit board that attaches to a Raspberry Pi computer, that does essentially the same thing as the DV-Thumb.

Tablets and Smartphones

I mentioned before that to my knowledge, as of this writing, there is no way to use either kind of D-Star dongle with an iOS or Android smartphone or tablet. The lack of a USB port makes it essentially impossible with an iPhone or iPad. There are hobbyists at work trying to get the Dongle to work with Android, but I've not heard of any breakthroughs yet.

One exception is with Microsoft Windows Tablets. The Microsoft Surface tablet is essentially a full Windows PC with all the usual ports on it. It can run the Dongle, DVAP, and DV-Thumb without any special attachments, drivers, or other hacks. It just works.

Another interesting option is the *very* inexpensive WinBook tablets that are out there. I have a TW700 tablet, made by Winbook, that I bought new at our local Micro Center for $59. Yes, that's a 7" tablet that runs Windows 8.1 for under $60. It's severely limited in storage, so it's not an outstanding computer, but it runs the DV-Dongle and DVTool software just fine. If you can find one of these tablets, or something similar, that's a fun way to have a dedicated D-Star machine. As an added bonus, the one I bought also included a one-year subscription to Microsoft Office, which normally runs $99 for a year.

USING D-STAR WITH THE ICOM IC-92AD

There are many excellent radios that support the D-Star system, and more new ones are coming to the market every few months. One very popular and very common radio over the past few years has been the Icom IC-92AD. It's a nice little handheld radio that, in addition to 2-meter and 440 bands, also does D-Star. It has a bright, clear display, and is relatively easy to program.

There are many accessories you can buy for the radio including a GPS microphone, fast chargers, and much more, but all you need to get D-Star going is just the basic radio--nothing else is absolutely required. That being said, I would *strongly recommend* the **RT Systems WCS-92 Programming kit** for the radio. The radio is fully programmable through the keypad, but it's tedious, error-prone, and not much fun. The WCS-92 makes the process much better. Another option is CHIRP, a free program that allows you to program a variety of radios. If you go with CHIRP, you'll still need a programming cable, and not all functions of the 92AD are supported. So again, I recommend just buying the programming software from RT Systems. The money is well spent in the time and hair-pulling it will save you.

The rest of this chapter assumes you have an Icom IC-92AD radio and the WCS-92 Software and cable. Icom makes many different radios, so even if you don't have this exact radio, or are using some other programming software, such as CHIRP, the ideas contained here will apply to most of the older Icom D-Star radios with a little adjustment. If you are programming the radio by using the keypad, that's fine too, just refer to the manual to explain how to access the programming functions.

Getting Started

If you skipped the "Essentials of D-Star" chapter, go back and work through it, as you will need to know the information for your repeater, frequency, offset, and three strings: YOUR, RPT1, and RPT2. Repeat this step as necessary to get connection information for **every repeater** you plan to program for the next step.

The next step is to program all this repeater information into the radio. If you have the RT Systems software, then enter the information you just got from the website into the tab (at the bottom) for "B and B Memories." D-Star on the IC-92AD only works from VFO B, so all your D-Star stuff must be programmed into the "B" memories. Why? It's a limitation of the radio.

Enter the frequencies, offset, offset direction, and "DV" for the Operating Mode. DV stands for Digital Voice, by the way. The main programming screen looks a lot like an Excel spreadsheet, and you navigate through it in much the same way as a spreadsheet program.

Important: Be careful when entering the RPT1 and RPT2 callsigns to **make sure you enter exactly the number of spaces that are indicated by the D-Star Calculator** (from the Essentials Chapter), since the format of the spacing matters. In my example from that chapter, the "blue box" instructions said RPT1 is "W8RTL**C" where the stars represent spaces. So I would carefully make sure I entered two spaces between W8RTL and the C when entering the callsign.

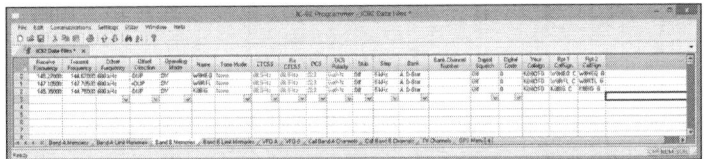

Again, enter information for every repeater you think you may want to use. You can always add more later. So if you already *know* which repeater is best, just one entry will be fine for now.

When you are finished with that step, pull down the "DStar" Menu, then choose the option "Digital (D-Star) Settings..." and you'll get a dialog box that looks like this:

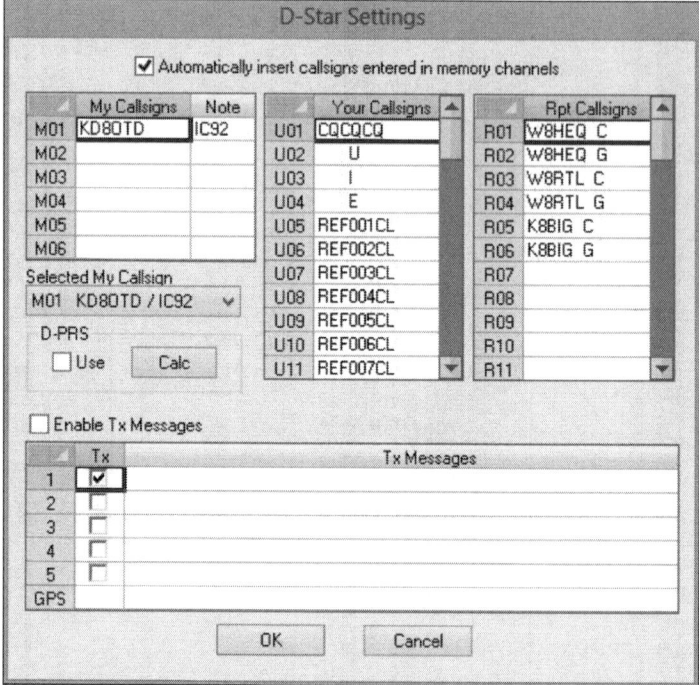

Set up the "My Callsigns" with your callsign. If more than one licensed ham will be using the radio, then enter all the applicable callsigns. If you are the only person who will use the radio, then one line is all you need.

Skipping over to the third box for a moment, "Rpt Callsigns," enter all the RPT1 and RPT 2 pairs that the D-Star calculator tool told you about. Depending on what you want to do, sometimes you'll need the "C" Connect function, and sometimes you'll want the "G" Gateway function. We should program in both of these for each repeater on our list.

I had three repeaters local to me, so I have six total entries in that box. You'll have the repeater name and either one or two spaces (depending on the length of the repeater callsign— you need eight characters in the whole field), and it ends with either C or G.

Programming the Destination

Reflector Linking

Now, for the complicated part. In the middle box, under "Your Callsigns," enter everything I have listed, at least for the first five lines. These aren't really callsigns; these are the various D-Star "command strings" that you will use to tell the D-Star repeater what to do. This includes things like "Connect to Reflector 1C" or "Unlink" or whatever you want the repeater to do. Enter them as I have them for convenience now. Notice that each line has exactly EIGHT characters. "REF001CL" and the others are easy, but note that "CQC-QCQ**" ends in two blank spaces and "*******U", "*******I", "*******E" all *begin* with seven spaces. You won't be able to see the spaces as you enter them, so count carefully.

I mostly talk on the reflectors, so, as you can see in the screenshot above, I have programmed mine to **Link** to module C on REF001, REF002, REF003, and so forth in cells U5 through the end of the list. You can put any individual callsigns, reflectors, or repeaters here that you want.

New reflectors get added to D-Star occasionally. There is a complete list of all the current ones as well as what they are primarily used for at:

http://www.dstarinfo.com/reflectors.aspx
Or take a look at **Appendix B**

Check out the list; maybe you know you won't use certain ones. I've added a large number of them to my system "just in case," but for the most part, I have found there are only a handful of reflectors that I actually use regularly.

I can't show all of them in screen capture above, but as of this writing, there are 71 reflectors. Just to make it even more complicated, your radio only has sixty memory slots, so you cannot simply enter them all. Take a look at the list in Appendix B or the link above and choose only the ones you think you might use. It's not hard to eliminate a bunch of them; I generally omit the ones for places that don't use English. You will probably want to program in regular repeaters or individuals as well as reflectors, so leave some blank memories for that.

Also, each reflector has an A, B, C, and sometimes D module. I usually only connect to the "C" modules. If you want to connect to the "A" and "B" modules, you'll need to program those in as well, just change the letter in the YOUR CALLSIGN string to "REF002AL" or something similar.

On the same screen, don't worry about D-PRS or enabling TX messages unless you want to experiment with those.

Repeater Linking

With D-Star, you also have the option to connect to other D-Star compatible repeaters. If you know you will be connecting to a repeater in another city, you can program in the callsign of the repeater here. For example, if I wanted to talk through the K8BIG repeater in Cincinnati, I would add a line (under "Your Callsigns") like this:

K8BIG*CL

Again the "*" is where you enter a space. This will (L)ink me to Module (C) of K8BIG.

Link to a Specific User

You can also use D-Star to connect to a specific, individual D-Star user in a long-distance conversation. Most of the setup is the same, but you just set the "Your Callsigns" field to the other person's callsign. If I wanted to call K8ABC and use my local W8RTL repeater, I'd set up a memory like this:
K8ABC***

Saving to the Radio

Now it's time to save your programming information to your radio. Make sure your radio is plugged into the computer, and assuming you are using the RT software, choose "Communication" from the main menu. From the Communication menu, choose "Send data to radio..." and follow the on-screen instructions from there. You should just need to turn your radio off and then on again, press the "OK" button, and wait a minute or two as your data is sent to the radio.

Once the radio has accepted all the programming, you are set to go. Turn off the radio, disconnect the cable, and then turn your radio back on. Assuming everything went well, you are ready to start using it.

Be aware that the RT Systems software is also very convenient for programming regular non-D-Star repeaters, simplex frequencies, and other VHF/UHF frequencies. I have mine set up for D-Star obviously, but also for all my local 2m and

440 repeaters, as well as setting up a separate memory "bank" that holds all the standard simplex frequencies. It makes choosing a frequency quick and convenient when you need one.

Using D-Star on the Now-Programmed Radio

1. Turn on the radio

2. Click the round "Main/Dual" button until VFO B is active. You can listen and talk in dual VFO mode, it doesn't really matter, but the "B" channel must be the active channel to talk or do anything with D-Star.

3. If you are not already in Memory Recall mode, press the "MR" button on the middle-right side of the numeric pad. A little "MR" symbol appears on the screen.

4. Turn the Control Dial until you see the call letters or frequency for your D-Star repeater. In the picture below, W8RTL is programmed in for memory location 001. You can also see the "B" indicator showing this is for VFO B and the DV indicating Digital Voice mode.

5. Click the "Menu" button once quickly to see the following:

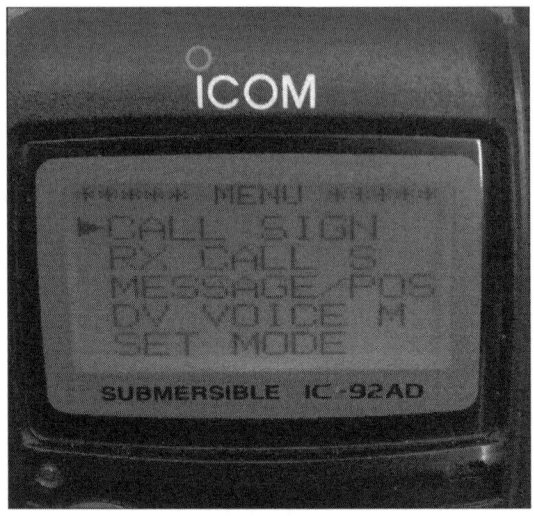

You will want to enter the "Callsign" menu option, so click the "5" button (for SELECT).

6. Now you'll see something like this:

R1 is the repeater you are connecting to (RPT1) through your local radio link. Generally this repeater must be within a few miles of your location. The "C" after the callsign means that you are connecting to Module "C" of that repeater. On the next line, R2 (RPT2), the "G" means that you are using that repeater as a "Gateway" to another system somewhere else. If you are going to link to either a reflector or distant repeater, then R2 is always set to "G." "MY" is simply my callsign, and "/IC92" is my radio type.

You probably won't be changing R1, R2, and MY very often unless you travel, but the UR field is where you choose your various command strings and destinations. In the picture above, I have REF001CL selected. This is the command to Link (L) to module C (C) on REF001. If for some reason you wanted to connect to module A or B, then you create a line back in the programming file that adds "REF001BL" to link to module B for example. As you can see, if you want all options for every possible reflector programmed in, this can get complicated quickly. It's made worse by the fact that there are only sixty memory slots.

To change the UR command string, click "5" for Select again from the above menu. You'll get something like this:

Now turn the control knob on the top of the radio. The various options for the UR field will change on the top line.

The one in the image above is "CQCQCQ," the standard setting for when you want to actually talk to people. When you're ready to speak, you want to use CQCQCQ mode.

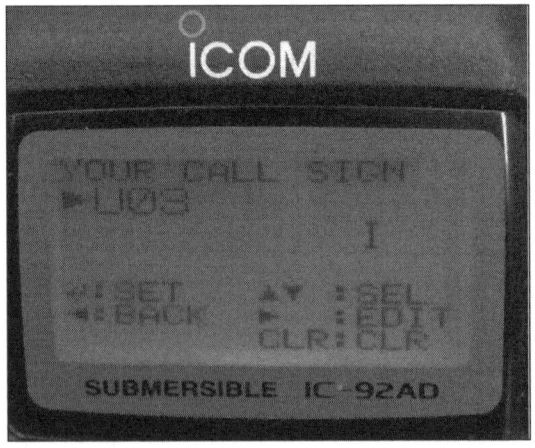

This one is command string "I" for information. Requesting information will return details about whether or not the repeater you are calling is already connected to anything or not.

This one is command "U" for Unlink. If someone else has connected to a reflector and you want to "change the channel," you should unlink first, then link to whatever you want. Make sure no one is using the repeater before you unlink.

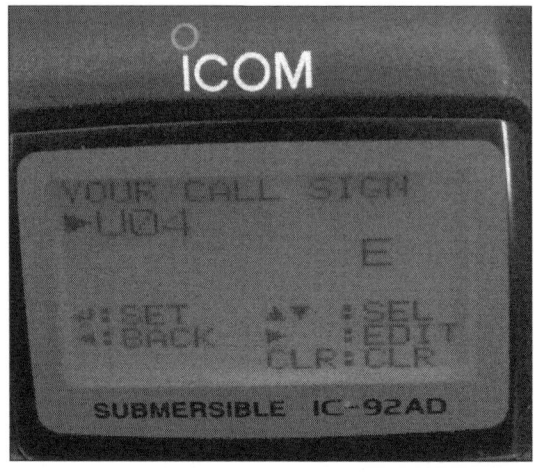

Callsign "E" is for "Echo." It's a way you can test your connection.

Turn it once more, and the various reflector or repeater connection commands will appear, starting with the first one:

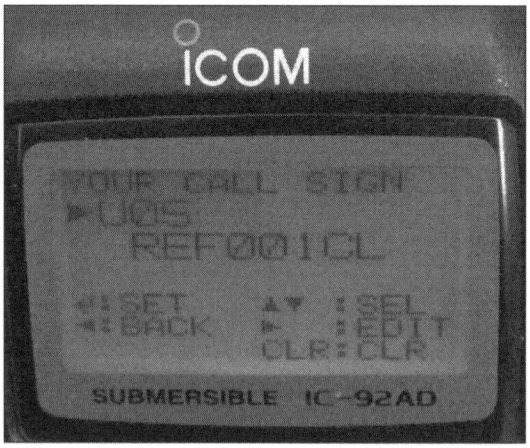

Followed by whatever other reflector and/or repeater connection strings that you entered.

Now let's actually *do* something with this. Turn the control knob backwards until the "I" (Information) command is selected. Press "5" (Select) to choose it. Now you'll see something like this screen:

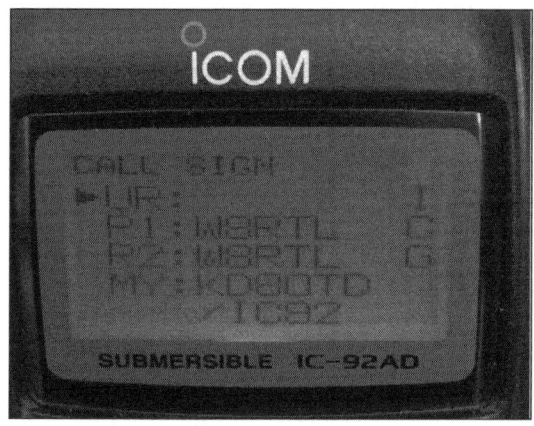

UR should be set to I

R1 and R2 should be set to your local repeater, and MY should be set to your own amateur radio callsign.

If it looks right, click the PTT button and hold for about a half-second. Depending on your local repeater, the radio's transmit light should briefly turn red (for transmitting) then green as it gets the information back from the repeater. Look at the readout on the bottom of your screen. I don't have a screen shot because it scrolls, but mine says "REF001 /C" that means my local repeater is already connected to Reflector 001, module C. My local repeater defaults to that reflector, since it's a pretty busy place most of the time.

Yours may very possibly say something else, or even "Module is unlinked," meaning the repeater isn't connected to anything right now. That's fine, it just means no one has used the repeater in a while, and it's not connected to anything.

Go back into the menus and choose the command string to connect to Reflector 001 module C. To recap:

1. Hit the menu Key
2. Select "Callsign"
3. Select "UR"
4. Turn the knob until you see "REF001CL" and select it.
5. Press the PTT button.

If all is well, you'll hear "Module is linked" or some similar message. Now go back through the above steps, and choose the "CQCQCQ" string in step 4. Press PTT and call CQ or ask for a radio check. From this point, it's just like talking on a local repeater.

If you've connected to a non-local repeater, another individual, or something other than a busy reflector, it is good etiquette to "Unlink" the repeater when you're done talking or listening. Someone else might assume the repeater was set to REF001, and it's really linked to somewhere far away. To unlink the repeater, just follow the steps on the previous page, but instead of the line where you chose "REF001CL," choose " U" (with seven spaces) to unlink.

If it worked, and you had a conversation with someone, then all is well. If not, you need to figure out why. Click on the PTT button again, say your callsign, and then quickly head over to **http://dstarusers.org**

If you are actually connected to the network, your callsign *should* appear near the top of the list, but if the D-Star network is really busy, you may need to scroll down a bit. If your callsign is on the list, then at least you got connected to the network; perhaps no one else was listening to answer your call. If it's not on the list at all, then there is either a problem on your end, or your local repeater could be down. Sometimes the whole reflector system stops working for a short period of time.

Double-check your command codes, repeater frequency, offset, and PL codes if your repeater needs one. Make sure you are in DV mode, and everything else. If all of that looks good, the problem could be with the repeater or reflector. It happens. Do you hear anyone else talking? Did the "I" (Information) command return anything at all? Are you actually in range of the repeater you are trying to use? There are a lot of factors that can go wrong, any one of which will keep you off the system. On the bright side, once you have it all set up and tested, it's actually pretty reliable.

Try a different local repeater if you have a choice. Try again in an hour. Give a general call on that repeater for assistance in setting up a D-Star connection. Perhaps there's something special about your specific repeater, like it's down for repair this week or something.

Look to your own settings first, but D-Star isn't perfect. Various repeaters and reflectors go down all the time.

USING D-STAR WITH THE ICOM IC-ID5100A

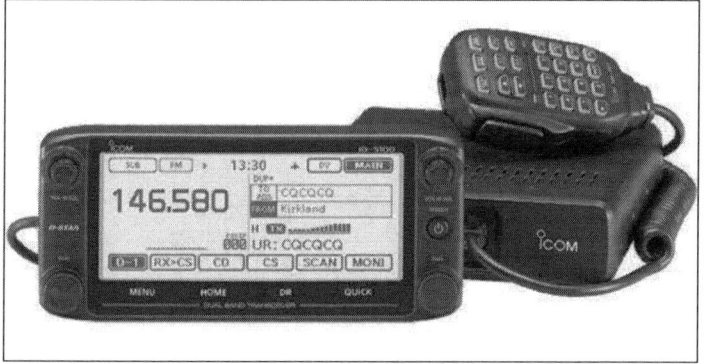

ICOM IC-ID5100A

The Icom IC-ID5100A is one of Icom's newest (mid-2014) D-Star compatible radios. It's very different to program and use than the older IC-92AD radio. The most visible difference is that the new mobile radio has a large touchscreen rather than a keypad. Although the touch screen is very nice for navigating menus and choosing options, what really makes the radio shine is that it has a huge database of repeaters, reflectors, and D-Star commands built right in, so a lot of the complexity of the previous chapter is no longer an issue. Want to link to a reflector? Just use the menus to do it. Want to enter somebody's callsign and call them directly? Just type it in on-screen. It's far less trouble, and in many cases, everything you need for D-Star is already programmed in there somewhere. In the rest of this chapter, we'll figure out where everything is, and hopefully figure out the best ways to make things happen on your ID5100A.

The small "quick-start" manual that comes with the radio is probably good enough to get you started. I would strongly recommend getting the PDF of the full manual. That can be downloaded here:

http://www.icom.co.jp/world/support/download/manual/

While we're at it, the 5100 comes with an excellent, worldwide database of D-Star and VHF/UFH repeaters. This list can be updated and Icom promises to make new versions of their databases available for download from:

http://www.icom.co.jp/world/support/download/firm/

Unfortunately, at the time of this writing (early 2015), the database on that link has the same date as the one included in the radio, so it's not necessary. If that changes, you should probably try updating the list of repeaters. I have found that the built-in list of D-Star repeaters includes everything in my local area, but I am not terribly impressed with the scope of the local VHF/UHF (non D-Star) repeaters. It lists *two* repeaters for Dayton, and I know there are at least a dozen active VHF repeaters in my area.

Getting Started

Assuming everything is downloaded, updated, and plugged in, let's skip straight to the D-Star part.

The first thing you should do is enter your callsign. It's easy:

1. Press the Menu button on the control panel and then [My Callsign] from the on-screen menu.

2. It will give you a blank list where you can enter more than one callsign if you need to, but for now, you'll press "1" to enter the first callsign.

3. Use the on-screen keyboard to type in your callsign.

4. Press the "ENT" on-screen button to enter your completed callsign.

5. You'll then be returned to the callsign list, where you should now see the callsign you just entered. Press on that callsign to set it as the one you want to actually use.

6. Press the Menu button again to return to the main display.

That looks like a lot of steps, but it's actually very simple; it's much like entering information on a smartphone. You will find that most functions and features use these menus and controls in exactly the same way.

Making a Simplex Call

You can make simplex calls (radio-to-radio direct calls) to other digital radio users, and this doesn't need a repeater.

1. Touch DR on the control panel.

2. Press the [From] button on the screen, and choose [Repeater List]. You aren't really using a repeater, but as you'll see on the next menu (called "Repeater Group"), one of the choices at the end of the list is [Simplex]. So click that.

3. Several frequencies will be displayed. Choose the one you want to use by turning the selection knob.

4. Press the "To" button on screen, and choose "Local CQ" The screen will now show "CQCQCQ" in the "To" field.
5. Press the PTT button on the microphone and talk.

Repeater Usage

With the radio's built-in database, setting up your radio for use with your local repeater couldn't be much easier.

Getting your local repeater set up:

1. Touch DR on the control panel

To automatically choose your local repeater:

2. Press the [From] button on the screen, and choose [Near Repeater]. This will bring up the "Near Repeater" menu. Click on [Near Repeater (DV)] for the digital voice repeaters near you. The radio will briefly show [Searching...] on screen as it uses its built-in GPS to determine your location, and then use that location to look up the repeaters that are near you. The database isn't huge, but hopefully it will show something you can reach.
3. Choose the repeater that is physically closest to you, and choose the Module (B or C) that you want to use. The repeater's frequency, offset, duplex, and so forth are automatically set for you!

OR... To manually choose your local repeater:

2. Press the [From] button on the screen, and choose [Repeater List]. This will bring up the "Repeater Group." Click on whichever region is closest to you.

3. Choose the repeater that is physically closest to you. The repeater's frequency, offset, duplex, and so forth are automatically set for you!

To use your local repeater to talk to local hams

Perform steps 1-3 as above to set up your local repeater

4. Touch the [To] field on-screen, then choose "Local CQ." The [To] field will display "CQCQCQ."

5. Press the PTT to connect and call CQ or otherwise make yourself known through that far-distant repeater.

To contact a distant repeater

Perform steps 1-3 as above to set up your local repeater

4. Touch the [To] field on-screen, then choose "Gateway CQ." The previous steps already set up your local repeater as your gateway, so all you need to do now is choose a destination. This brings up the "Repeater Group" list of regions. Choose the region you want from the list, and this will bring up a list of repeaters in the countries or cities within that region. Choose one.

The database that ships with the 5100A is somewhat limited. It has a lot of D-Star repeaters, but new ones are being added all the time. If you are trying to call a distant repeater, you can choose [Direct Input (RPT)] from the "To" menu and enter it in manually.

5. Press the PTT to connect and call CQ or otherwise make yourself known through that far-distant repeater.

Calling a Specific Individual

Perform steps 1-3 as above to set up your local repeater

4. Touch the [To] field on-screen, then choose "Direct Input (UR)."
5. Enter the callsign of the individual you wish to contact.
5. Press the PTT to connect and call his sign or otherwise make yourself known.

Link to and use a reflector

Perform steps 1-3 as above to set up your local repeater

4. Touch the [To] field on-screen, then choose "Reflector." After this you'll have the choice of

[Use Reflector]
[Link Reflector]
[Unlink Reflector]
[Echo Test]
[Repeater Information]

5. Choose "Link Reflector." As you use the reflector, you will return to this menu to choose various options later.
6. When you are first starting out, your screen will only show "Direct Input." After you have connected to a few reflectors,

your favorites will be added to this screen. For now, choose "Direct Input."

7. Choose the REF number (Reflector) by number, using the + and - keys to change the number. You can also use the second set of + and - keys to choose which Module you want (A through Z, although only A-D are currently common). After you've set the Reflector number and module letter, press "ENT" to enter the value.

8. Press the PTT to connect and link to the Reflector.

9. If it worked, you should hear a message stating "Linked to REF001 Charlie" or something similar.

10. Click on the "To" field again, and choose "Reflector" once again. This time, choose "Use Reflector." You will be returned to the main screen, where you should see "Use Reflector CQCQCQ" in the "To" field.

To get information about your local reflector (like whether or not it's currently linked or where it's linked to), go back to step 5 and choose "Repeater Information." Select that option and click the PTT for a second. Watch your screen for a quick status report. Unlinking the reflector is done the same way; just choose the appropriate option in step 5.

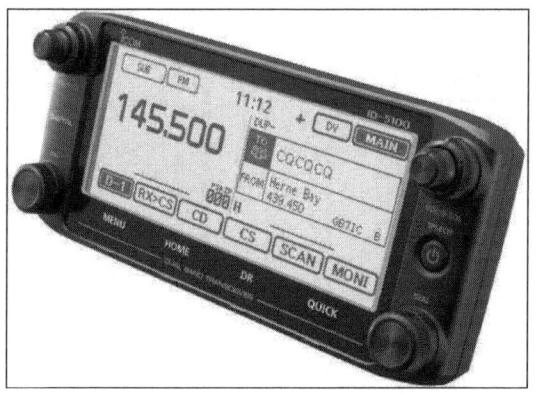

USING D-STAR WITH THE DVAP

DVAP

If you don't have or don't want to buy a D-Star radio, then you will need the DV-Dongle to get on the D-Star network. But what about people who have a compatible radio but don't have a local D-Star repeater? That's where the DVAP comes in.

The DVAP is a dongle similar to the DV-Dongle but with a few differences. The DVAP has a little antenna that sticks out of the end. Otherwise, the only visible difference is the color: The DV-Dongle is blue, while the DVAP is red.

The"Red Dongle" - the DVAP

Basically, when using a DVAP, you plug the device into your Internet-ready computer. You use a D-Star compatible radio, either handheld or mobile, to connect to a frequency that you set on the DVAP. The DVAP then converts your radio signal to data files and sends them through the Internet, in very much the same way the DV-Dongle does with your voice. The primary difference is that the DV-Dongle uses your computer's speaker and microphone, while the DVAP lets you carry around a radio for the same task. It's like having your own private repeater in your home.

The range on the little DVAP antenna is only about 100 meters, so you can carry your HT radio around the house and yard, but it's not meant to be used as a "public repeater."

Once you are registered on the D-Star network (See Appendix A) you are ready to go. Here are the steps to setting it up:

1. Plug in the DVAP. If you are running Windows or Linux, then skip to step 2. For Mac OSX users, you'll need to download a special Virtual Com Port driver to let OSX see the DVAP: **http://www.ftdichip.com/Drivers/VCP.htm**
2. The next step in getting the DVAP to work is to download the software. It can be found at:

http://www.opendstar.org/tools/

The file you will want is the latest version of DVAPTOOL offered. As of this writing, it's at version 1.04, and there are different versions for Windows, OSX, and Linux. Download and install the latest version for your system. Watch the filenames carefully, there are similar-looking files there named DVTOOL, but those are for the DV-Dongle, not the DVAP.

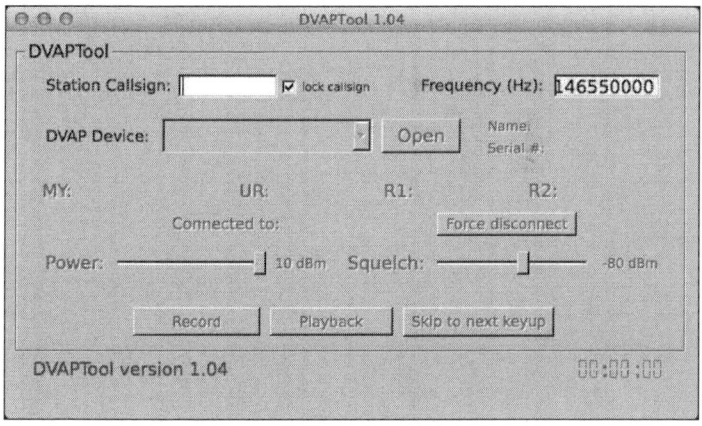

3. Enter your amateur radio callsign in the callsign box. It is probably a good idea to go back to the website of the gateway where you registered (See Appendix A) and add a terminal entry with an "A" in the initial field. Then enter your

callsign with the "A" in the eighth character in DVAPTool. The "lock callsign" checkbox is used to prevent others from transmitting using your DVAP. If you want to allow other licensed amateur operators to use your DVAP, then uncheck the box.

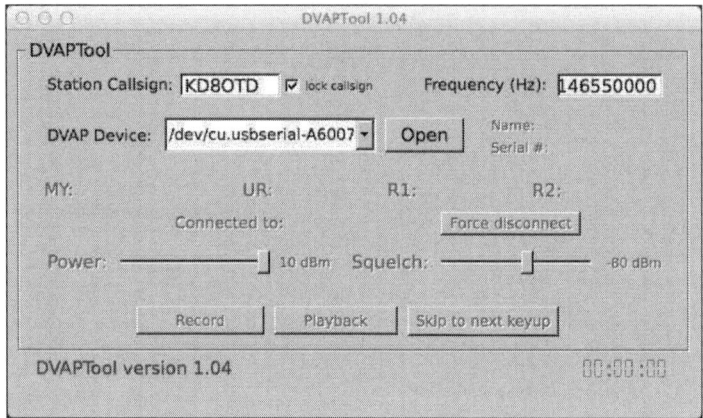

4) Enter a 2-meter simplex frequency for operation. Enter the entire frequency in Hz (e.g. 146550000 for 146.550 MHz).

What Frequency Do I Use?

The most important consideration when choosing a frequency is obviously whatever is legal for your country, and you'll need to consult a band plan to answer that question.

For example, within the USA, 2-meter simplex frequencies are 146.400 through 146.595. The national calling frequency is 146.52, so you wouldn't want to use that one.

The ARRL's USA Band plan is at:

http://www.arrl.org/files/file/Regulatory/Band%20Chart/Hambands_color.pdf

Another good 2-meter reference for USA hams:

http://www.konr.com/rwitte/2m_frequencies.html

For other countries, just search for your allocated frequencies and eliminate anything that is reserved for other uses.

The DVAP itself puts out very little power, only around ten milliwatts, so it has almost no chance of causing interference with other hams, but your handheld radio's transmit power should be lowered to the minimum that will work. A half-watt to a single watt of output power will usually be more than enough to reach the DVAP within the short distance that it can return the connection.

5) Select the appropriate "DVAP Device" from the drop-down list and then click "Open." From this point forward, all the control commands are issued from your D-Star radio. The only other feature of the computer software that you might find interesting is the ability to record and play back audio files.

That's all there is to setting up the computer side of the DVAP.

Setting up the commands on the radio is very similar to that explained in the previous chapters with the following differences:

Enter your callsign into the MYCALL field of your radio. Select DV mode and configure simplex operation, usually by holding down the "DUP" key until there is no "dup+" or "dup-" on the radio display. You do not need to enter RPT1 or

RPT2 since the radio will place "DIRECT" in both when in simplex mode. All commands are entered into the URCALL field as detailed below.

Set the UR: field to an eight-character command string just as before. Again, the "*" represents filler-spaces, and you need to make sure the correct number of spaces are entered or else the command won't work:

"DVAP*I"** (request Information from the DVAP)
"DVAP*E"** (press PTT and speak for Echo test)
"CQCQCQ"** (transmit to a connected system)
"xxxxxxmL" (to link, replace XXXXXX with the gateway or reflector callsign making sure to use 6 characters filling the end with spaces as needed. Replace the "m" in the 7th character with the module to which you wish to link. Use "L" in the 8th character to indicate the Link command. For example, to link to Reflector 001 module C, use "REF001CL" in URCALL. To link to gateway W4DOC module A, use "W4DOC*AL" in URCALL.
"*****U"** (unlink from a linked gateway/reflector)

As you can see, the setup is mostly the same, but a few of the command strings need to be altered.

Keeping these differences in mind, the majority of the radio set up and usage is just like that shown the previous chapters.

RASPBERRY PI INTERFACING

Raspberry Pi Main Board

Over the past couple of years, the Raspberry Pi miniature computer has exploded in popularity with hams and computer hobbyists. A Raspberry Pi is a tiny little device that is essentially a full computer, easily interfaced with other devices. If you don't want to tie up your main computer with your DVAP, you can plug them into a tiny, inexpensive Raspberry Pi computer and have the whole thing be an easily portable, dedicated system. With the DVAP, you can "set-it and forget-it" since it won't tie up or rely on your regular PC.

This book isn't about the Raspberry Pi, and the topic is relatively complex for the beginner, so I'm just going to include a few sources to get you started on your research if the topic interests you:

DVAP and Raspberry Pi How to:
http://www.dstar101.com/DVAPrpi.htm

Youtube Videos:
How to Setup a Raspberry PI and DVAP for D Star:
https://www.youtube.com/watch?v=FKjHw3xVnsw

DV-MEGA Raspberry Pi Radio:
https://www.youtube.com/watch?v=kHDnEOwqOXo

Portable DVAP station / Raspberry PI / Cellular Hotspot / VNC: https://www.youtube.com/watch?v=7r7VwU-NK1A

GOING FORWARD WITH D-STAR

Here are a handful of additional resources and things to get involved with using D-Star. These may or may not appeal to everyone, but if you want to do more with D-Star than I've described so far, these various sites should be your next step.

D-Star nets:

http://www.dstarinfo.com/nets.aspx

Nets are fun ways to listen to users around the world (or just locally) talk about specific topics. There is usually some kind of net on one of the reflectors each night.

D-Star reflectors

http://www.dstarinfo.com/reflectors.aspx

Reflectors are the "Chat Rooms" of D-Star. Some of them, such as Reflector 1 (REF001C) usually have quite a few people listening, and if you give a CQ, someone will almost always answer. Some of the reflectors are pretty specific, so look at the list and make a note of the ones you are most likely to be interested in.

See who else is on D-Star and where they are:

http://www.d-starusers.org/

This link is a good way to see if there are any "busy spots" on the D-Star network. It's also a quick way to test your radio: key the microphone for a half a second (if no one's talking) and see if your callsign shows up on the list. If it doesn't, you might not be connected to the net properly.

Computer Data Tools (D-Rats):

http://www.d-rats.com/

D-RATS is a communications tool for D-STAR low-speed data (DV mode). It provides:
Multi-user chat capabilities
File transfers
Structured data transport (forms)
Position tracking and mapping.

Other D-Star Tools:

http://www.dstarinfo.com/applications.aspx

Other interesting projects and links:

For more advanced callsign routing, take a look at the IRCDDB project: http://IRCDDB.net/live.htm

If you want a little more advanced version of the software for DV-DONGLE or DVAP, then give WIN-DV a try. WIN-DV software / Dutch-Star: http://dutch-star.eu

APPENDIX A: D-STAR SELF-REGISTRATION

1. Open a web browser and enter **http://www.dstarusers.org/repeaters.php** to find the D-Star repeater system closest to you. Click on the link for your repeater. Here is the page for my local repeater (W8RTL):

```
                        System Information
                        Callsign: W8RTL
                           City: Dayton
                           State: Ohio
                         Country: USA
                         Website: http://www.w8bi.org
        Gateway Registration URL: https://w8heq.dstargateway.org/Dstar.do
                 Gateway Enabled: YES
              DSTARMonitor Enabled: YES
                     ARRL Listed: NO
                    Trust Server: US ROOT

                       Frequency Information

          2 Meters (Usually "C" Node): 147.10500MHz +0.600

       70 Centimeters (Usually "B" Node): 443.05000MHz +5.000

       23 Centimeters Voice (Usually "A" Node): 1283.50000MHz -12.000

           23 Centimeters Digital Data 1249.00000MHz

                       Additional Information
       This D-STAR repeater is located at Educational TV Channel 16
       transmitter site in Dayton, Ohio, and is maintained by the Dayton
       Amateur Radio Association. Antennas are located on east side of
       tower at the following heights above ground level: 2-meter - 5dBd,
       Celwave, PD-220 Super Stationmasters - Receive at 850', Transmit at
            610'. 70 cm - 10 dBd Celwave, PD-455 Super Stationmasters -
           Receive at 880', Transmit at 635'. 23 cm - 12 dBd Hustler, HS12-
                 12430 - Receive and Transmit at 610'.
```

Look for a link labeled "Gateway Registration URL" for your repeater. If a "Gateway Registration URL" is not listed for your closest repeater, contact the administrator of the repeater for instructions. Save or print out this page. You may need the frequency and offset information later when programming your radio. If you are only using the DV-Dongle, don't worry about saving this info.

2. You should see the web page shown below. Click **"Register"** to begin the New User registration process.

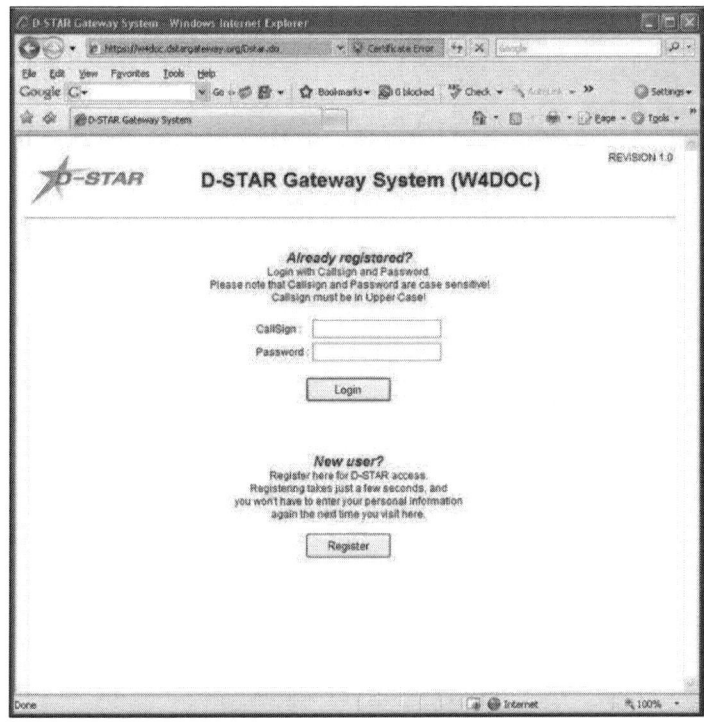

3. You should see the web page shown below:

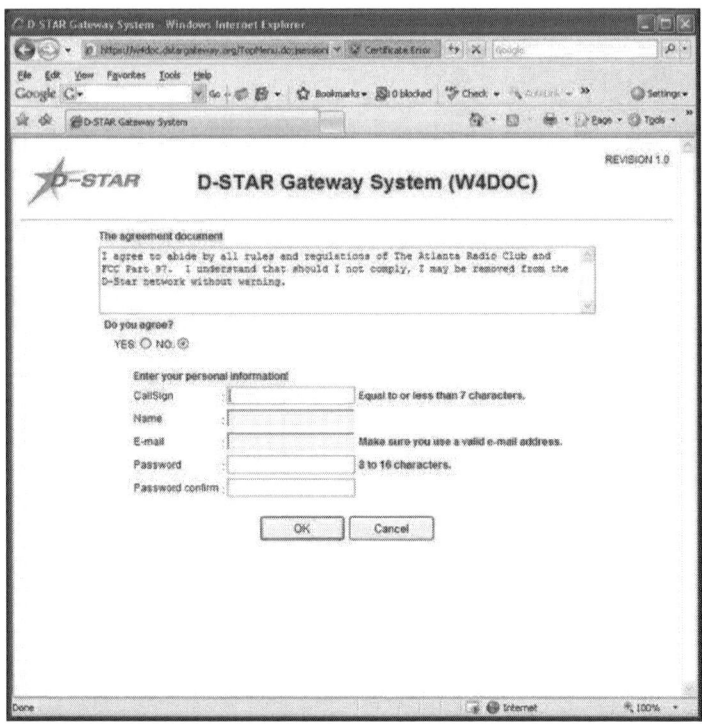

4. Click on "Yes" to agree to the registration terms. Enter your callsign in the Callsign field IN UPPER CASE. Enter your full name in mixed case, and your email address. Enter your new password in both fields. Passwords must be at least 8 characters in length and no longer than 16. Click "OK" when ready to continue.

5. The following popup window should be displayed. Click "OK".

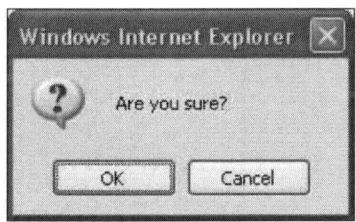

6. The following web page should be displayed. Click "OK." Your registration request has been submitted for approval. The gateway administrator must now log into the system and approve your request. The approval may happen quickly or it may take a day or two depending on how often the administrator logs into the system.

7. You should try to log in periodically to check on the status of your registration. Use the same Gateway Registration URL you found in step 1, and use your callsign (IN UPPER CASE) and the password you entered during registration to attempt to log in. If your registration is **still pending** approval, you will see the following web page.

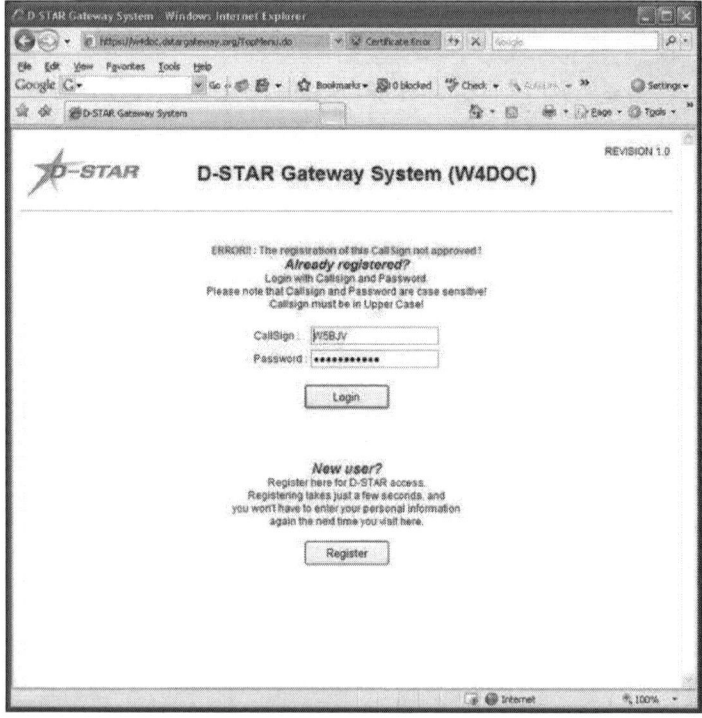

8. Once your registration has been approved, you will be able to log into the system and configure your personal information. THIS STEP IS REQUIRED. Once you are logged in, click on "Personal Information" at the right of the page. You should see a web page similar to the one in step 9.

9. Click on the checkbox next to the number "1". Then click inside the "Initial" box to the right of your callsign on the same line as the number "1." Type in a single-space character. This will not show up but is very important. Do not click on the "RPT" check box. In the "pcname" box, enter your callsign in lower case followed by a dash "-" followed by your type of radio, e.g. 2820 or DV-Dongle. All characters in the "pcname" box should be lower case and there should be no spaces. Your web page should look similar to the following page (with your personal information shown). When complete, click on "Update."

10. Once you have clicked "Update," you should see the following popup box.

11. Click "OK" and your registration is COMPLETE! You may log out and begin using the D-Star network.

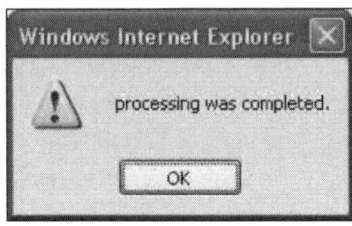

Welcome to D-Star!!!

APPENDIX B: LIST OF REFLECTORS

Reflector - Usage - Location

REF001A London, England

REF001B Illinois D-STAR repeaters London, England

REF001C D-STAR's MegaRepeater London, England

REF002A Southeastern US D-STAR Weather Net NE, USA

REF002B Some Nets NE, USA

REF002C Some Nets NE, USA

REF003A Ad-hock & Emergency Use - Australia Australia

REF003B Permalink for Repeaters, including all WIA Port B Repeaters — Australia Australia

REF003C Australian Nets Australia

REF004A Alternate for Southeastern US D-STAR Weather Net USA

REF004B Texas Permalink Repeaters USA

REF004C General Rag Chew (English only please) USA

REF005A UK Nets, Permalink Repeaters London, England

REF005B Kent Net (UK Repeaters around Kent) London, England

REF005C London, England

REF005D UKFMGW Net (North West UK Repeaters) London, England

REF006A Scottish Net London, England

REF006B London, England

REF006C German Net London, England
REF007A Florida Orlando, FL, USA
REF007B Florida Orlando, FL, USA
REF007C Florida Orlando, FL, USA
REF008A Japan G2 repeaters, DV Dongles and DVAPs Japan
REF008B Japan G2 repeaters, DV Dongles and DVAPs Japan
REF008C Japan G2 repeaters, DV Dongles and DVAPs Japan
REF009A AZ, USA
REF009B AZ, USA
REF009C Arizona Permalink Repeaters AZ, USA
REF010A Emergency Communications New England, USA
REF010B Open New England, USA
REF010C New England Repeaters New England, USA
REF011A Italy
REF011B Italy
REF011C Italy
REF012A Permalink Repeaters Southern California, USA
REF012B Southern California, USA
REF012C Southern California, USA
REF013A London, England
REF013B London, England
REF013C London, England
REF014A US west coast repeater linking NE, USA
REF014B US west coast repeater linking NE, USA
REF014C US west coast repeater linking NE, USA
REF015A Multimedia (non-DSTAR) London, England
REF015B Multimedia (non-DSTAR) London, England
REF015C Data Only - Worldwide use London, England
REF016A British Columbia, Canada
REF016B British Columbia, Canada
REF016C British Columbia, Canada
REF017A Netherlands (Dutch Speaking repeaters, hotspots and dongles) Amsterdam, the Netherlands
REF017B Amsterdam, the Netherlands

REF017C Amsterdam, the Netherlands
REF018A Brazil USA
REF018B Brazil USA
REF018C Brazil USA
REF019A WI, USA
REF019B WI, USA
REF019C WI, USA
REF020A NJ, USA
REF020B NJ, USA
REF020C NJ, USA
REF021A Wales and the West Country London, England
REF021B Midstar / Midlands London, England
REF021C London, England
REF022A Nordic region - Finnish Espoo, Finland
REF022B Nordic region - Swedish/Danish/Norwegian Espoo, Finland
REF022C Nordic region - English/other activity Espoo, Finland
REF023A Australia
REF023B Australia
REF023C Australia
REF024A Michigan ARES USA
REF024B Indiana Permalink Repeaters USA
REF024C General Use USA
REF025A Public Service, Skywarn & Emergency Use Washington, DC, USA
REF025B National Capital Region Association Permalink Washington, DC, USA
REF025C General Usage & Sunday Night NCR Net Washington, DC, USA
REF026A General Use Vancouver, BC, Canada
REF026B British Columbia D-Star Repeaters Vancouver, BC, Canada
REF026C Provincial Emergency Radio Communications Vancouver, BC, Canada
REF027A Italy Italy
REF027B Italy Italy
REF027C Italy Italy

REF028A Belgium (Flemish Speaking Belgian repeaters, hotspots and dongles in Flanders area) Diest, Belgium
REF028B Belgium (French Speaking Belgian repeaters, hotspots and dongles in Wallonia area) Diest, Belgium
REF028C Belgium (Slow Data experiments like D-RATS,D-STAR TV...) Diest, Belgium
REF029A Northern Utah General Usage and Linking UT, USA
REF029B Northern Utah General Usage and Linking UT, USA
REF029C Northern Utah General Usage and Linking UT, USA
REF030A Georgia Special event use Atlanta, GA, USA
REF030B Georgia Linked Repeaters (no out of state permalinking) Atlanta, GA, USA
REF030C Wide area linked repeaters Atlanta, GA, USA
REF030D Georgia digital data use (D-RATS) Atlanta, GA, USA
REF031A Sweden
REF031B Sweden
REF031C Sweden
REF032A Poland
REF032B Poland
REF032C Poland
REF033B Dallas, TX, USA
REF033C Dallas, TX, USA
REF034A Florida Jacksonville, FL, USA
REF034B Florida Jacksonville, FL, USA
REF034C Florida Jacksonville, FL, USA
REF035A Washington State WA. USA
REF035B Washington State WA, USA
REF035C Washington State WA, USA
REF036A Midstar UK Worcester, United Kingdom
REF036B Wales and West Country Groups Worcester, United Kingdom
REF036C Dongle To Dongle and Raynet Worcester, United Kingdom
REF037A Florida Orlando, FL, USA
REF037B Florida Orlando, FL, USA
REF037C Florida Repeaters Orlando, FL, USA

REF038A The Ohio Reflector Cleveland, OH, USA
REF038B NODIG Permalink Cleveland, OH, USA
REF038C Dayton Hamvention and the ARRL Cleveland, OH, USA
REF039A Ohio State Wide ARES Wellington, OH, USA
REF039B Ohio State Wide EMA Wellington, OH, USA
REF039C Ohio State Wide Severe Weather Wellington, OH, USA
REF039D American Red Cross Operation in Ohio and surrounding Wellington, OH, USA
REF040A Portugal Lisbon, Portugal
REF040B Portugal Lisbon, Portugal
REF040C Portugal Lisbon, Portugal
REF041A Chicago, OH, USA
REF041B Chicago, OH, USA
REF041C Chicago, OH, USA
REF042A Prague, Czech Republic
REF042B Prague, Czech Republic
REF042C Prague, Czech Republic
REF043A Gothenburg, Sweden
REF043B Gothenburg, Sweden
REF043C Gothenburg, Sweden
REF044A Seattle, WA, USA
REF044B Seattle, WA, USA
REF044C Seattle, WA, USA
REF045A Emergency Calls Only (Worldwide Use) Athens, Greece
REF045B Data Calls Only (Worldwide Use) Athens, Greece
REF045C Voice Calls Only (Worldwide Use - Primarily Greek Language) Athens, Greece
REF046A Florida Orlando, FL, USA
REF046B Florida Orlando, FL, USA
REF046C Florida ARES use Orlando, FL, USA
REF047A Japan G2 repeaters, DV Dongles and DVAPs Japan
REF047B Japan G2 repeaters, DV Dongles and DVAPs Japan
REF047C Japan G2 repeaters, DV Dongles and DVAPs Japan
REF048A Louisiana Statewide Ultimate D-STAR Shreveport, LA, USA

```
REF048B Louisiana Statewide Ultimate D-STAR Shreveport, LA, USA
REF048C Louisiana Statewide Ultimate D-STAR Shreveport, LA, USA
REF049A North East US General Use/Nets New York, NY, USA
REF049B North East US Repeater Linking New York, NY, USA
REF049C North East US Emergency/Skywarn New York, NY, USA
REF049D North East US Data New York, NY, USA
REF050A New England Repeaters Boston, MA, USA
REF050B New England Repeaters Boston, MA, USA
REF050C New England Repeaters Boston, MA, USA
REF051A Emergency Communications Chicago, IL, USA
REF051B Club and other non-emergency nets Chicago, IL, USA
REF051C Casual Use / Ragchewing Chicago, IL, USA
REF052A Oklahoma Moreland, Oklahoma, USA
REF052B Oklahoma Moreland, Oklahoma, USA
REF052C Oklahoma Moreland, Oklahoma, USA
REF053A General use Minneapolis, MN, USA
REF053B Nets and general use Minneapolis, MN, USA
REF053C SKYWARN, ARES/RACES, public service, nets and general use
Minneapolis, MN, USA
REF054A Carolinas General Use Charlotte, NC, USA
REF054B Carolinas EmComm Use / ARES Charlotte, NC, USA
REF054C Carolinas Repeaters Charlotte, NC, USA
REF055A New Mexico repeaters and surrounding states New
Mexico, USA
REF055B New Mexico ARES, RACES, S&R, Skywarn New Mexico, USA
REF055C General Use New Mexico, USA
REF055D Data / D-RATS New Mexico, USA
REF056A Kentucky General Use KY, USA
REF056B Kentucky General Use KY, USA
REF056C Kentucky General Use KY, USA
REF056D Kentucky Digital ONLY KY, USA
REF058A Alabama General Use Auburn, AL, USA
REF058B Alabama Nets and Emergency Communications Auburn, AL, USA
REF058C Alabama General Use Auburn, AL, USA
```

REF059A Missouri repeaters /EmComm St. Louis, MO, USA
REF059B Missouri repeaters /EmComm St. Louis, MO, USA
REF059C Missouri repeaters /EmComm St. Louis, MO, USA
REF060A Tennessee EmComm/ARES Use Nashville, TN, USA
REF060B Tennessee/Southeast repeaters Nashville, TN, USA
REF060C Tennessee/Southeast repeaters Nashville, TN, USA
REF060D Tennessee digital data use (D-RATS) Nashville, TN, USA
REF061A South Carolina ARES/EmComm only Florence, SC, USA
REF061B South Carolina Repeater Linking & ARES/EmComm Florence, SC, USA
REF061C South Carolina Linked Repeaters Florence, SC, USA
REF061D South Carolina data use Florence, SC, USA
REF062A North Central / Northeast Maryland ACS Repeaters Hunt Valley, MD, USA
REF062B North Central / Northeast Maryland ACS Nets Hunt Valley, MD, USA
REF062C Maryland ACS Linked Repeaters Hunt Valley, MD, USA
REF062D Maryland ACS Digital Data Hunt Valley, MD, USA
REF063A Pittsburgh, PA, USA
REF063B Pittsburgh, PA, USA
REF063C Western Pennsylvania Repeaters Pittsburgh, PA, USA
REF064A Near Narita Airport Repeaters Chiba, Japan
REF064B Near Narita Airport Repeaters Chiba, Japan
REF064C Near Narita Airport Repeaters Chiba, Japan
REF065A Worldwide HF DX Voice Spotting Asheville, NC USA
REF065B HF DXing Chat Asheville, NC, USA
REF065C North Carolina Linked Repeaters Asheville, NC, USA
REF065D North Carolina data use Asheville, NC, USA
REF066A K5ELK, Elk City, Oklahoma Dallas, Texas, USA
REF066B General Use Dallas, Texas, USA
REF066C D-STAR on US Route 66 Dallas, Texas, USA
REF066D General Use Dallas, Texas, USA
REF067A HamWAN discussion and activities Memphis, Tennessee, USA
REF067B Mid-South Nets Memphis, Tennessee, USA

```
REF067C General use Memphis, Tennessee, USA
REF067D Digital data use (D-RATS) Memphis, Tennessee, USA
REF068A Italy Italy
REF068B Italy Italy
REF068C Italy Italy
REF068D Italy Italy
REF069A Emergency Use ONLY CT, USA
REF069B General Use CT, USA
REF069C CT DSTAR Group Northeast Network Permalink CT, USA
REF069D Digital Data Use or other testing CT, USA
REF071A Western part of Japan Yamaguchi, Japan
REF071B Western part of Japan Yamaguchi, Japan
REF071C Western part of Japan Yamaguchi, Japan
REF071D Western part of Japan Yamaguchi, Japan
```

ABOUT THE AUTHOR

Brian Schell (KD8OTD) is a former College IT Instructor who has an extensive background in computers dating back to the 1980s. Currently, he writes on a wide array of topics from computers, to world religions, to ham radio, and even releases the occasional short horror tale.

He'd love to hear your stories of success and failure with D-Star. If there's something you would like to see in a future edition of the book, or otherwise have suggestions, please drop him a note. Contact him at:

```
Web: http://BrianSchell.com
Email: brian@brianschell.com
```

twitter.com/BrianSchell

facebook.com/Brian.Schell

instagram.com/brian_schell

pinterest.com/brianschell

ALSO BY BRIAN SCHELL

Amateur Radio
- D-Star for Beginners
- Echolink for Beginners
- DMR for Beginners Using the Tytera MD-380
- SDR for Beginners with the SDRPlay
- OpenSpot for Beginners
- Programming Amateur Radios with CHIRP

Technology
- Going Chromebook: Living in the Cloud
- Going Text: Mastering the Power of the Command Line
- Going iPad: Ditching the Desktop
- DOS Today: Running Vintage MS-DOS Games and Apps on a Modern Computer

The Five-Minute Buddhist Series
- The Five-Minute Buddhist
- The Five-Minute Buddhist Returns
- The Five-Minute Buddhist Meditates
- The Five-Minute Buddhist's Quick Start Guide to Buddhism
- Teaching and Learning in Japan: An English Teacher Abroad

Fiction with Kevin L. Knights:
- Tales to Make You Shiver

- Tales to Make You Shiver 2
- Random Acts of Cloning
- Jess and the Monsters

Made in the USA
Monee, IL
13 July 2021